Industry 5.0

Industry 5.0 is a new paradigm aiming to address social and environmental sustainability in future industries. This book explores various models and execution frameworks across different sectors, enabling organizations to stay competitive by integrating digital technologies with human creativity. It is a resource for smart manufacturing and service organizations, offering insights and strategies for next-generation industrial practices.

Industry 5.0: Concepts and Strategies for Digital Transformation offers a comprehensive toolkit designed to enhance organizational sustainability within the Industry 5.0 landscape. This resource provides in-depth insights for practitioners and researchers, presenting a range of implementation frameworks to improve sustainability practices across industries. The book explores the key drivers and technological advancements driving Industry 5.0 while also addressing the unique barriers faced by different industrial sectors in adopting this transformative approach.

This book provides practitioners and industry leaders with the tools to effectively address contemporary challenges associated with product life cycles and societal impacts. It also helps organizations enhance their resilience to achieve the objectives outlined in the Paris Agreement and COP 28. Furthermore, it assists policymakers and government agencies in formulating emission control policies and mitigation strategies.

Sustainable Manufacturing Technologies: Additive, Subtractive, and Hybrid

Series Editors: Chander Prakash, Sunpreet Singh, Seeram Ramakrishna, and Linda Yongling Wu

This book series offers the reader comprehensive insights of recent research breakthroughs in additive, subtractive, and hybrid technologies while emphasizing their sustainability aspects. Sustainability has become an integral part of all manufacturing enterprises to provide various techno-social pathways toward developing environmental friendly manufacturing practices. It has also been found that numerous manufacturing firms are still reluctant to upgrade their conventional practices to sophisticated sustainable approaches. Therefore this new book series is aimed to provide a globalized platform to share innovative manufacturing mythologies and technologies. The books will encourage the eminent issues of the conventional and non-conventional manufacturing technologies and cover recent innovations.

Manufacturing Engineering and Materials Science
Tools and Applications
Edited by Abhineet Saini, B. S. Pabla, Chander Prakash, Gurmohan Singh, Alokesh Pramanik

Manufacturing Technologies and Production Systems
Principles and Practices
Edited by Abhineet Saini, B. S. Pabla, Chander Prakash, Gurmohan Singh, Alokesh Pramanik

Modern Hybrid Machining and Super Finishing Processes
Technology and Applications
Edited by Ankit Sharma, Amrinder S. Uppal, Bhargav P. Prajwal, Atul Babbar and Chander Prakesh

Green Manufacturing and Materials Processing Methods
Characterizations, Applications, and Design
Edited by Sarbjeet Kaushal, Sandeep Bansal, Chander Prakash, Bhupinder Singh and Dheeraj Gupta

Advanced Manufacturing Processes
Integration of Sustainable Approaches
Edited by Hitesh Vasudev, Chander Prakash, Mohammad Khalid, and Kavindra Kesari

Industry 5.0
Concepts and Strategies for Digital Transformation
Edited by Mahender Singh Kaswan, Rajeev Rathi and Jose Arturo Garza-Reyes

For more information on this series, please visit: https://www.routledge.com/Sustainable-Manufacturing-Technologies-Additive-Subtractive-and-Hybrid/book-series/CRCSMTASH

Industry 5.0

Concepts and Strategies for Digital Transformation

Edited by
Mahender Singh Kaswan, Rajeev Rathi, and
Jose Arturo Garza-Reyes

CRC Press
Taylor & Francis Group
Boca Raton London New York

CRC Press is an imprint of the
Taylor & Francis Group, an informa business

Designed cover image: Summit Art Creations: Shutterstock

First edition published 2025
by CRC Press
2385 NW Executive Center Drive, Suite 320, Boca Raton FL 33431

and by CRC Press
4 Park Square, Milton Park, Abingdon, Oxon, OX14 4RN

CRC Press is an imprint of Taylor & Francis Group, LLC

ISBN: 978-1-032-87821-8 (hbk)
ISBN: 978-1-032-87886-7 (pbk)
ISBN: 978-1-003-53518-8 (ebk)

DOI: 10.1201/9781003535188

Typeset in Times
by KnowledgeWorks Global Ltd.

Contents

Preface

The book provides a significant knowledge base about the different concepts associated with the new industry paradigm Industry 5.0. It provides a detailed description of the potential technologies associated with Industry 5.0 that lead to improved operational excellence by keeping in mind mother nature. Furthermore, the book also demonstrates key enablers and potential barriers to the adoption of Industry 5.0. The book also highlights the key application areas of Industry 5.0 that will enhance the scope of this Industrial Revolution. It also symbolizes how the different industrial sectors including education, agriculture, manufacturing, and fashion can include different aspects of Industry 5.0. Furthermore, the book also fosters the relationship between human intelligence and key Industry 5.0 technologies that will further boost the prospect of Industry 5.0 for improved organizational sustainability. The book also demonstrates the integration of Lean philosophy with Industry 5.0 that will boost the prospect of integration among different operational excellence approaches and Industry 5.0 technologies.

About the Editors

Mahender Singh Kaswan is working as an associate professor in the Department of Mechanical Engineering, Lovely Professional University, Phagwara, Punjab, India. He received his Ph.D. degree from Lovely Professional University, Phagwara, Punjab, India in 2021 and M.Tech. from the National Institute of Technology (NIT), Kurukshetra, India. His research areas include Industry 4.0/5.0, Circular Economy/NET Zero, Lean Six Sigma, Operation Management, Green Technology, and Sustainable Human Resource Management. He has published several articles in reputed journals like the *Journal of Cleaner Production*, *IEEE Transactions on Engineering and Management*, *International Journal of Lean Six Sigma*, and *Environmental Impact Assessment and Review*. He is also serving as a guest editor in *Engineering Management Journal* (SCI Q1), *Management of Environmental Quality* (SCI Q1), and *TQM Journal* (SCOPUS Q1). He is also a member of different reputed organizations like the Indian Science Congress. He has qualified for the Graduate Aptitude Test in Engineering (GATE) a slew of times. He has published 55 publications, five patents published, and one guaranteed.

Rajeev Rathi is an assistant professor (senior grade) in the Department of Mechanical Engineering, National Institute of Technology, Kurukshetra, India. His research is mainly focused on operational excellence and clean technologies in the manufacturing and service sectors. He is currently working on the development and implementation of various operational excellence models, mainly based on Lean, Six Sigma, Lean Six Sigma, and Kaizen. He is also working on sustainability, social sustainability dynamics in the industrial scenario, and Industry 4.0. He has published over four patents, four granted patents, and 62 articles in peer-reviewed international journals and conference proceedings. He is an active guest editor, editorial board member, and reviewer of many high-repute journals. He is working in collaboration with international researchers from various reputed universities such as Heriot-Watt University, UK; Khalifa University, Abu Dhabi; University of Derby, UK; Texas Technical University, USA; University of Dayton, USA; National University of Ireland Galway, UK; and Namibia University of Science and Technology, Namibia, in the field of operations excellence, Lean Six Sigma, Lean, sustainability, Industry 4.0, and process improvement in manufacturing and service.

Jose Arturo Garza-Reyes is a professor of Operations Management and Head of the Centre for Supply Chain Improvement at the University of Derby, UK. He is actively involved in industrial projects where he combines his knowledge, expertise, and industrial experience in Operations Management to help organizations achieve excellence in their internal functions and supply chains. He has also led and managed international research projects funded by the British Academy, British Council, European Commission, and Mexico's National Council of Science and Technology (CONACYT). As a leading academic, he has published over 250 articles in leading scientific journals, at international conferences, and seven books. Prof. Garza-Reyes is Associate Editor of the *International Journal of Operations and Production Management*, Associate Editor of the *Journal of Manufacturing Technology Management*, Editor of the *International Journal of Supply Chain and Operations Resilience*, and Editor-in-Chief of the *International Journal of Industrial Engineering and Operations Management.* Areas of expertise and interest for Professor Garza-Reyes include general aspects of operations and manufacturing management, business excellence, quality improvement, and performance measurement.

Contributors

Andrei Bonamigo
Federal Fluminense University – UFF
Brazil

Jessica Borg
School of Management at Royal Melbourne Institute of Technology University
Melbourne, VIC, Australia

Davide Calandra
University of Turin
Turin, Italy

Abdelkabir Charkaoui
Faculty of Sciences and Technique, Hassan First University of Settat
Morocco

Rekha Chaudhary
Lovely Professional University
Phagwara, Punjab, India

Anass Cherrafi
EST-SAFI, Cadi Ayyad University
Marrakech – Safi, Morocco

Chidambara
School of Planning and Architecture
Delhi, India

Anindita Das
ASTHA School of Management
Bhubaneswar, Odisha, India

José Eugênio Monteiro de Camargo Filho
Federal Fluminense University – UFF
Brazil

David de Oliveira Costa
University of Pernambuco – UPE
Brazil

Raphael Odebrecht de Souza
Santa Catarina State University (UDESC)
Florianópolis, Santa Catarina, Brazil

Khadija Echefaj
Faculty of Sciences and Technique, Hassan First University of Settat
Morocco

Sana Elhidaoui
UCA Marrakech
Morocco

Hajar Fatorachian
Leeds business School, Leeds Beckett University
Leeds, UK

Fernando Antônio Forcellini
Santa Catarina State University (UDESC)
Florianópolis, Santa Catarina, Brazil

Rekha Goyat
Lovely Professional University
Phagwara, Punjab, India

Upasna Goyal
National Institute of Technology
Kurukshetra, Haryana, India

Mohammad Rafid Irtisham
Aston University
Birmingham, United Kingdom

Rejaul Karim
Varendra University
Rajshahi, Bangladesh

Mahender Singh Kaswan
Lovely Professional University
Phagwara, Punjab, India

Ish Kumar
School of Planning and Architecture
Delhi, India

Rakesh Kumar
Lovely Professional University
Phagwara, Punjab, India

Sudhansu Sekhar Nanda
Sri Sri University
Cuttack, Odisha, India

Rajeev Rathi
National Institute of Technology
Kurukshetra, Haryana, India

Md. Mustaqim Roshid
Rajshahi University
Rajshahi, Bangladesh

José Carlos Sá
Polytechnic of Porto.
Porto, Portugal

Geeta Sachdeva
National Institute of Technology
Kurukshetra, Haryana, India

Keith Heringer dos Santos
Federal Center for Technological Education Celso Suckow da Fonseca – CEFET
Brazil

Mahipal Singh
Lovely Professional University
Phagwara, Punjab, India

Abdul Waaje
Michigan Technological University
Houghton, MI

Steffan Macali Werner
Federal Santa Catarina University – UFSC
Brazil

1 Industry 5.0 and Its Different Facets

Mahender Singh Kaswan, Rakesh Kumar, and Rekha Chaudhary

1.1 INTRODUCTION

Industry 5.0 (I5.0) represents a paradigm shift in the way we conceptualize and implement industrial and manufacturing processes. Building upon the advancements of its predecessors, I5.0 aims to seamlessly integrate human intelligence and capabilities with cutting-edge technologies such as artificial intelligence (AI), cobots, digital twins, zero defect manufacturing, big data, and the internet of things (IoT) [1, 2]. However, to unlock the full potential of I5.0 and create a truly interconnected and intelligent ecosystem, several key enablers must be in place. Now, it is a well-known fact that implementation always remains a challenge, whether it is a process or a policy, and I5.0 is no exception. Many articles have identified the obstacles and barriers to successful implementation of I5.0; however, study of its enablers has been very limited [3–5]. This chapter explores and investigates the enablers of I5.0 in order to improve organizational performance. The process of transformation from I4.0 to I5.0 is undoubtedly cumbersome for many organizations in terms of infrastructure, training, and managerial decision-making because the new industrial paradigm emerged rapidly after I4.0. Therefore, an in-depth study of enablers will play a vital role in the success of I5.0.

1.2 KNOW-HOW OF INDUSTRY 5.0

I5.0 improves organizational sustainability through logical integration of a smart set of technologies and machines and through continuous monitoring powered with human creativity. Further, I5.0 fosters production quality by assigning repetitive tasks to robots and creative thinking tasks to human beings [6, 7]. I5.0 development made a considerable leap forward when the European Commission launched its agenda for future industry under the banner of I5.0. The commission strongly argued that current model, I4.0, was not best suited to the prevailing climate and social conditions.

I5.0 differs from I4.0 in the following respects:

- I5.0 induces technical development by enabling humanoids using human-focused approaches.
- I5.0 encourages stakeholders' predominance in the governance of technology, innovation growth, and management of social enactment.
- I5.0 prompts both productivity-driven effectiveness and ecological development.

DOI: 10.1201/9781003535188-1

- I5.0 is based on certain technologies and principles that enhance corporate social sustainability aspects across the entire value chain of the product.
- I5.0 prompts expertise-focused innovative practices within the ambit of environmental sustainability.

1.3 DIFFERENT INDUSTRIAL PARADIGMS

Industrial organizations are in continuous pursuit of improved sustainability dynamics through the adoption of ecofriendly methods and practices. In the past few decades, increased use of traditional methods of operation and increased utilization of natural resources have led to a substantial increase in the earth's temperature [8]. This has increased the level of greenhouse gas emissions, which consequently has resulted in health-related issues in the workplace and the surrounding community. The history of I5.0 can be traced back through the development of different industrial revolutions [9, 10]. Traces of I1.0 can be found in the 1780s, when mechanical power was being generated from fossil fuels, steam, and water. The first industrial revolution was about mechanical production, as the horse and human power were replaced by steam engines. The second industrial revolution occurred between the late 1800s and the start of World War I. I2.0 was more about electricity, production of steel, and combustion engines. The concept of mass production was introduced in I2.0, and this revolution was also termed the revolution of electrification. I3.0 was more focused on use of computers, the internet, and digitization. This third revolution was also called the digital revolution and can be traced back to around the 1980s. This revolution saw the automation of production utilizing electronic and IT systems. I4.0 started in initial years of 21st century. Much of I5.0 goes beyond I4.0 and uses human expertise in conjunction with smart, efficient, and intelligent machines to avoid delay and dissonance, make the workplace safer and industry more environmental friendly, and improve organizational productivity [11].

1.4 INDUSTRY 5.0 AND SUSTAINABILITY

I5.0 prompts sustainable development by meeting key sustainable development goals like socioeconomic sustainability, resilience building, and the centrality of humanoids (Ghobakhloo et al., 2022). In the past few decades, the academic literature and policy debates have focused on sustainability and sustainable development. In the literature, there are different reasons for adopting I5.0 in an effort to improve social and environmental sustainability. I5.0 can address the limitations of I4.0 to create a human-centered system that improves ecofriendly sustainability measures (Özdemir & Hekim, 2018). It is also reported that I5.0 prompts an improved working ambiance, greater employability, and increased industrial efficacy (Kumar et al., 2021). The development of I5.0 took a substantial leap forward when the European Commission revealed the future perspective of European technology leaders under the umbrella of I5.0. The commission reported that I5.0 is a humanoid-focused and ecologically valued system and must embrace novel technology directed toward socioenvironmental developments. I5.0 can be envisaged as an extension of I4.0 to prioritize socioeconomic needs. It makes the foundation shift from linear economy models that affect environmental sustainability to a smart manufacturing paradigm (Aslam et al., 2020). I5.0 fosters flexible manufacturing design and manufacturing using a cognitive cyber-physical system that in turn enables interacted systems of information and production to deliver highly customized products (Özdemir & Hekim, 2018). I5.0 technologies foster better risk management, resource planning, and resource reconfiguration. Thus, it can be deduced from the literature that I5.0 contributes toward socioeconomic sustainability, coupled with economic sustainability, through better resource planning using a human-centered integrated system of smart machines, devices, and technologies and cognitive human abilities.

1.5 BARRIERS TO INDUSTRY 5.0

Barriers are those obstacles those hinder an organization when adopting a new business model or technology. Being a new industrial revolution, I5.0 also faces these obstacles. It is hindered by different barriers related to different aspects of the organization. For instance, when adopting I5.0, the first and foremost challenge is resistance to change, as organization members always resist adopting a new model. Furthermore, I5.0 adoption is challenged by the lack of funds, particularly in the context of small and medium enterprises in developing nations. I5.0 is also restricted by other barriers related to technology, integration across the value chain, and technological and human thinking interfaces.

Barrier 'financial constraints' can be mitigated by developing financial institutions that provide easier credit access to the industries that are looking to adopt I5.0 practices. Government policies that encourage easier access to bank loans can enable industries to procure technologies and update infrastructure needed to adopt I5.0. Recently, different fiscal policies in India provided easier credit access at a subsidized rate. This benefited the country's manufacturing firms by enabling them to adopt advanced technologies for the production of high-precision, mass-customized products. Furthermore, to overcome barriers related to resistive culture, organizations must provide education and training on the different facets and technologies of I5.0. To boost the practices related to sustainability, industries should forge ties with organizations that prompt sustainability initiatives, with academia, and with other research- and innovation-focused institutions to upgrade technical knowledge and capacity through different training activities.

1.6 CONCLUSION, LIMITATIONS, AND FUTURE ASPECTS

I5.0 provides a new set of advanced technologies coupled with the human intelligence that is the future for all industrial sectors. I5.0 makes industry more sustainable by keeping humans at the center of all technologies. It has found application in different industrial sectors like manufacturing, medicine, agriculture, shipping, and education. To cope up with these new challenges, industrial organizations must adopt I5.0 technologies and also cognitive human thinking to prevent defects and accidents and to provide more value by keeping a focus on human beings and environmental sustainability. Future researchers can find ways to interlink different operational management approaches with I5.0, to develop a practical model for implementing I5.0, and to integrate the circular aspects of the economy with I5.0.

REFERENCES

1. P. K. R. Maddikunta *et al.*, "Industry 5.0: A survey on enabling technologies and potential applications," *Journal of Industrial Information Integration*, vol. 26, Mar. 2022. doi: 10.1016/j.jii.2021.100257.
2. J. Leng *et al.*, "Industry 5.0: Prospect and retrospect," *Journal of Manufacturing Systems*, vol. 65, pp. 279–295, Oct. 2022. doi: 10.1016/j.jmsy.2022.09.017.
3. S. Grabowska, S. Saniuk, and B. Gajdzik, "Industry 5.0: Improving humanization and sustainability of Industry 4.0," *Scientometrics*, vol. 127, no. 6, pp. 3117–3144, Jun. 2022. doi: 10.1007/s11192-022-04370-1.
4. Skobelev, P. O., & Borovik, S. Y. (2017). *On the way from Industry 4.0 to Industry 5.0: From digital manufacturing to digital society.* Industry 4.0, 2(6), 307-311.
5. M. C. Zizic, M. Mladineo, N. Gjeldum, and L. Celent, "From Industry 4.0 towards Industry 5.0: A review and analysis of paradigm shift for the people, organization and technology," *Energies*, vol. 15, no. 14, Jul. 2022. doi: 10.3390/en15145221.
6. M. S. Kaswan, R. Chaudhary, J. A. Garza-Reyes, and A. Singh, "A review of Industry 5.0: From key facets to a conceptual implementation framework," *International Journal of Quality & Reliability Management*, ahead of print, 2024. doi: 10.1108/IJQRM-01-2024-0030.
7. U. Kumar, M. S. Kaswan, R. Kumar, R. Chaudhary, J. A. Garza-Reyes, R. Rathi, and R. Joshi, "A systematic review of Industry 5.0 from main aspects to the execution status," *TQM Journal*, vol. 36, no. 6, pp. 1526–1549, Jun. 2024. doi: 10.1108/TQM-06-2023-0183.
8. S. Y. Teng, M. Touš, W. D. Leong, B. S. How, H. L. Lam, and V. Máša, "Recent advances on industrial data-driven energy savings: Digital twins and infrastructures," *Renewable and Sustainable Energy Reviews*, vol. 135, Jan. 2021. doi: 10.1016/j.rser.2020.110208.
9. L. Gualtieri, E. Rauch, and R. Vidoni, "Emerging research fields in safety and ergonomics in industrial collaborative robotics: A systematic literature review," *Robotics and Computer-Integrated Manufacturing*, vol. 67, Feb. 2021. doi: 10.1016/j.rcim.2020.101998.
10. K. A. Demir, G. Döven, and B. Sezen, "Industry 5.0 and human-robot co-working," *Procedia Computer Science*, vol. 158, pp. 688–695, 2019. doi: 10.1016/j.procs.2019.09.104.
11. A. Correia Simões, A. Lucas Soares, and A. C. Barros, "Factors influencing the intention of managers to adopt collaborative robots (cobots) in manufacturing organizations," *Journal of Engineering and Technology Management*, vol. 57, Jul. 2020. doi: 10.1016/j.jengtecman.2020.101574.

2 Industry 5.0
A Human-Machine Collaboration: Approach and Applications

Anindita Das and Sudhansu Sekhar Nanda

2.1 INTRODUCTION

The rise of Industry 5.0 promises to revolutionise manufacturing through robot-human collaboration. Human labour and ubiquitous robots are boosting industrial productivity (Longo et al., 2020). Executive teams in industrial companies outline production lines, monitor KPIs, and ensure efficiency. Industry 5.0 is transforming production robots and industrial automation. Business productivity and production speed are rising due to AI and cognitive computing (Pathak et al., 2021). Industry 5.0 creates an eco-friendly, renewable-energy-powered system to encourage sustainability and production. Businesses must adopt Industry 5.0 by training personnel in AI and robotics to ensure robots and operators operate together (Leone et al., 2020). Businesses matter in cutting-edge tech decisions. Virtual education reduces training costs by eliminating work interruptions (Majumdar, et al., 2021). This strategy ensures safe training for personnel. Interactive classrooms improve staff collaboration and morale. Automation and AI are becoming more significant in the workplace.

2.2 INDUSTRIAL REVOLUTION PHASES

With the first industrial revolution, or Industry 1.0, the 18th century changed drastically. Machines displaced humans and outdated manufacturing methods. The United States followed England in this transition, which began around 1780. Mining, textiles, agriculture, glassmaking, and others were affected by Industry 1.0's shift from handicrafts to technology. Between 1871 and 1914, Industry 2.0 permitted faster movement and innovation in the industrial sector. Because machines replaced humans in manufacturing, the unemployment rate rose even if the economy and company productivity grew. In the 1970s, computers and memory-programmable controls launched Industry 3.0, the digital revolution. Mass production, digital logic, integrated circuit chips, computers, digital cellular phones, and the internet were used to digitalise items and company processes. Industry 4.0, which combines physical assets with AI, the IoT, robotics, 3D printing, and cloud computing, may help organisations adapt and make data-driven decisions. Next-generation technology, "Industry 5.0," creates smarter, more efficient equipment. Michael Rada's 2015

DOI: 10.1201/9781003535188-2

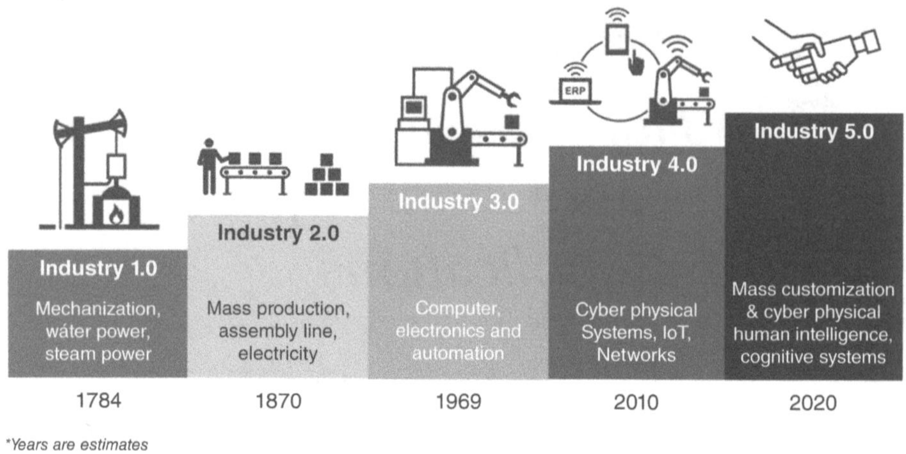

FIGURE 2.1 Industrial revolutions.

suggestion highlighted the importance of maximising human-robot collaboration in harmony with nature. Rada says it is a revolutionary change that will promote Industrial Upcycling, human-machine cooperation, and massive waste reduction. Figure 2.1 denotes different industrial revolutions.

2.3 PRINCIPLES OF INDUSTRY 5.0

Industry 5.0 is guided by three key principles—Sustainability, Human-centredness, and Resilience—that define its structure and goals. Below is a detailed explanation of each principle:

2.3.1 Sustainability

Industrial processes' environmental impact was largely disregarded in past industrial revolutions. Business executives have shown environmental concern through CSR programmes, but these efforts have had little impact. The fifth industrial revolution emphasises sustainable business methods over CSR. Companies must change their strategies to prioritise sustainability. Companies should prioritise the 17 Sustainable Development Goals (SDGs) and the Triple Bottom Line (TBL)—People, Planet, and Profit. Many corporations have focused on greenwashing or harm reduction till recently. True sustainability strategies boost a company's positive influence. Companies should improve the environment rather than just reduce their impact.

2.3.2 Resilience

COVID-19 impacted the global supply chain, producing shortages. This made a strong supply chain a priority for industries. Industry 5.0 intends to establish a resilient supply chain to maintain crucial supplies amid emergencies. Industries need resilience to face uncertainty. Manufacturing processes must be optimised for durability. To perform well during crises, companies must become nimbler and more flexible.

2.3.3 Human Centric

Industrial Revolution 5.0 also emphasises seeing human resources as assets rather than just tools. This human-centric approach puts people above organisations. Manufacturing companies should pursue workforce-enhancing technologies. Technology should adapt to workers' demands, not vice versa. A human-centric strategy prioritises technology that improves human roles in production. Workers should be empowered, not replaced by machines. Robots perform repetitive jobs efficiently and consistently, but they lack human problem-solving skills. Thus, the industrial industry is adopting a hybrid approach that combines machines and humans.

2.4 INDUSTRY 5.0 CREATIVE APPLICATIONS

2.4.1 Smart Healthcare

A smart hospital that processes patient data in real time is an Industry 5.0 aim. This technique enables healthcare remote monitoring. The doctors' life is substantially enhanced. This smart healthcare technology can assist physicians identify COVID-19 patients and provide them with the information they need to treat them better during the pandemic. It also helps medical students train during the COVID-19 pandemic. ML is used in genetic data, medical imaging, and NLP. Disease detection, prediction, and diagnosis are priorities. Figure 2.2 denotes different applications of Industry 5.0.

Abdelmageed and Zayed said Industry 5.0 allows for customised smart implants that meet consumer needs. Medical professionals are using AI systems for many purposes, including blood sugar monitoring. This allows mass customisation of orthopaedic implants to meet each patient's needs. Patient implants are made differently, and medical equipment can be upgraded. Technology is boosting medical education, learning, and research while increasing surgical precision. Industry 5.0 emphasises orthopaedics' need for high-quality, customised implants with long lifespans by addressing overproduction, tool selection, and transparency issues.

2.4.2 Manufacturing Industry

Industry 5.0, the fifth industrial revolution, reimagines manufacturing through human-machine cooperation. Human creativity and ever-improving technology should work better together.

Industry 5.0 recycles and reuses resources to reduce waste and protect the environment. Additive manufacturing is essential for personalisation, efficiency, and waste reduction. With this trend, automated systems are replacing human workers in many manufacturing processes worldwide. Intelligent systems and robotics are more prevalent in supply chains and manufacturing than ever, say Brown and Wobst. Smart manufacturing lets designers safely upload manufacturing item design files to the cloud, improving resource utilisation and access

Ghobakhloo et al. state that designers can locate factories wherever as long as they are near raw material suppliers and low-cost production zones. Cloud manufacturing manages these facilities' machinery and output. In this service-oriented strategy, production and services are integrated to create customised client solutions. To increase efficiency, value, and market share, businesses include service components

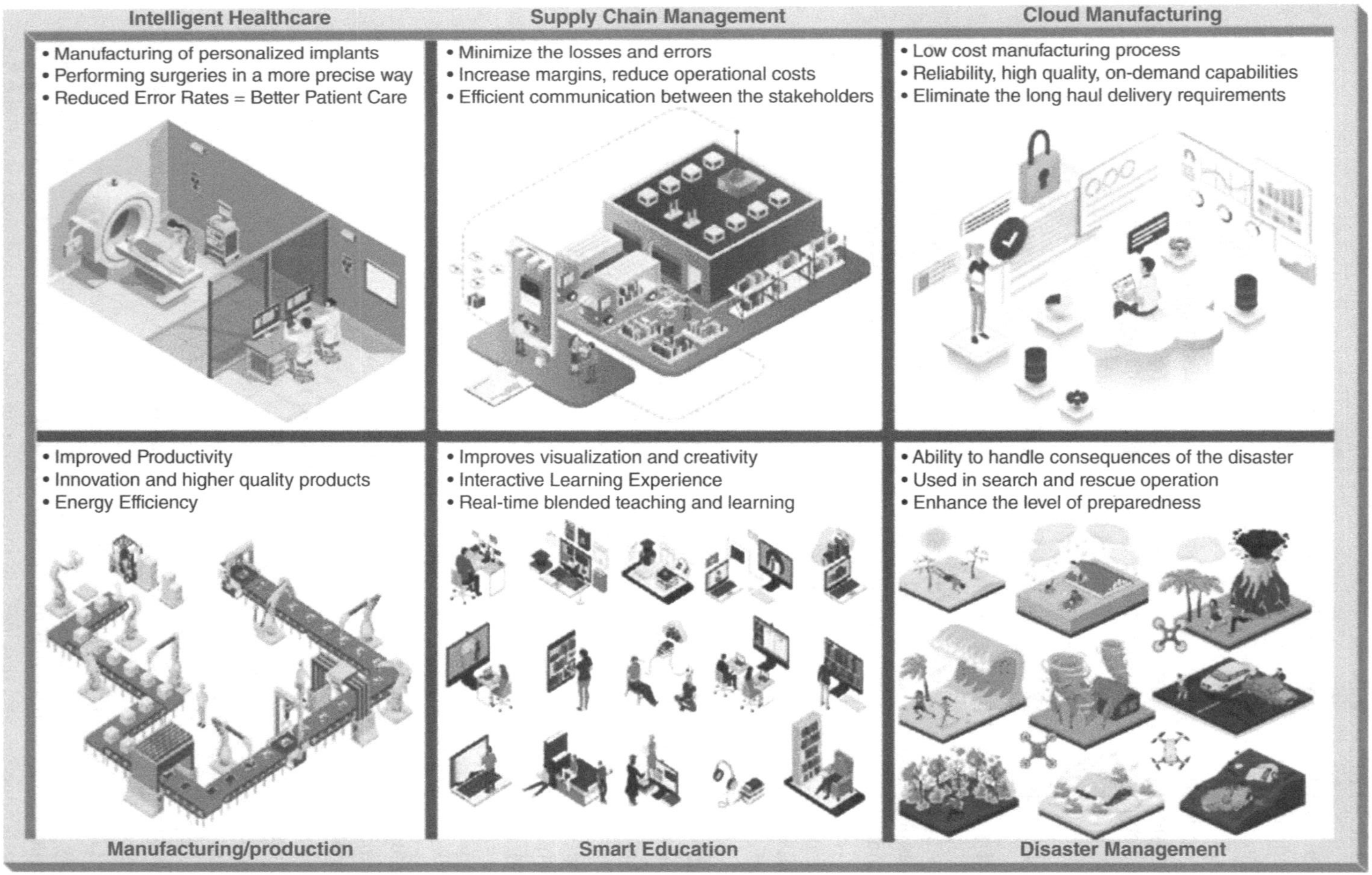

FIGURE 2.2 Industry 5.0 applications.

in production. As a distributed and networked production resource system, the cloud-based platform efficiently and cheaply manages manufacturing services.

2.5 TECHNOLOGIES USED IN INDUSTRY 5.0

Industry 5.0 is supported by enabling technologies such as cloud computing, blockchain, big data analytics, the Internet of Things (IoT), and 6G networks.

Cloud computing provides databases, software, analytics, and networks. Internet-accessible servers store and manage data for efficient innovation and cost-effectiveness. This technology offers on-demand applications, storage, and computing power. Cloud service providers have developed mobile and web-based IoT monitoring solutions as part of the industrial cloud's push to virtualise industry applications. Edge computing devices need limited resources for analytics, while APIs normalise data from numerous sources. In key sectors, worldwide cooperative cloud ecosystems are common.

Haleem, Javaid, and Khan say cloud computing can scale edge device data processing. Autonomous manufacturing robots use this infrastructure for IoT platforms. Industries access local server data daily for managerial purposes. Industry 5.0 aims to reduce centralised server data. Cloud computing enables preventive data analysis to spot machine issues early. This speeds up problem-solving and lets operations continue with extra people.

Industry 5.0 wants humans back in design and manufacturing. It lets humans use robots' precision and power to lift heavy objects. Excelling at key tasks allows humans to precisely control and personalise manufacturing phases. Industry 5.0 and collaborative robotics emphasise human participation in process improvement, ushering in a new era in manufacturing and robotics. Industry 5.0 emphasises personalised smart gadgets, cars, and other products, while Industry 4.0 emphasised quality uniformity and data collection. Industry 5.0 uses cutting-edge technologies like the IoT to boost human output with robots. Industry 4.0's automation and efficiency become humanised with robotics. IoT networks sensors and software-equipped physical objects to exchange data over the internet. Ghobakhloo et al. argue that the IoT ushers in a new age by addressing data security problems and modernising traditional lifestyles with cutting-edge technology. Sinclair et al. found that IoT security requires robust identity and access control solutions. Authentication systems prevent man-in-the-middle attacks and protect critical data during communications. Authentication and access control systems for cloud-based data-centric applications aim to maintain authenticity, confidentiality, and end-to-end latency in IoT communication networks.

Industry 5.0 technology uses three-dimensional symmetrical ecosystem designs. Big data analytics, according to Matheus et al., involves using huge databases to uncover market trends and patterns. It applies advanced analytical methods on structured and semi-structured datasets, requiring a lot of data storage and processing power. Predictive advice from real-time data analysis gives organisations an edge. Big data analytics aids strategic decision-making and identifies and fixes inconsistencies. We consider population demographics and location accessibility to determine customer preferences and optimise customer experiences. This strengthens customer ties. In Industry 5.0, the lack of production cycle data causes issues.

In blockchain technology, digital ledger records, or "blocks," store transaction data. This shared ledger lets businesses track transactions and assets. Business data is crucial, thus blockchain technology provides a shared, comprehensive, and irreversible ledger for all network participants. This aids order, payment, manufacturing, and other tracking. Distributed ledger technology allows network participants to track transactions without database duplication. Smart contracts on the blockchain automate transactions based on predefined circumstances, such as paid travel insurance. Verifying earlier blocks enhances a blockchain's transaction history, which is organised into irreversible chains. Businesses need precise data to authenticate transactions, and distributed ledgers make data sharing easier and save time.

The 6G cellular data network supports wireless communication. 6G companies will likely be more diverse. They might enable VR/AR, widespread instant messaging, and deep IoT interaction beyond mobile devices. 6G mobile network operators are expected to use decentralised, adaptive business models such as intelligent automated management, infrastructure sharing, local licencing, spectrum sharing, blockchain, and short-packet communication. Industry 5.0's 6G networks should match intelligent information society's high-reliability needs. Using artificial intelligence, mobility forecasting methods maintain network connectivity. High data rates for numerous applications challenge Industry 5.0. Energy management in Industry 5.0 is complicated by the growing number of connected huge smart devices. Reduce energy use and deploy energy harvesting systems to optimise energy management.

2.6 CONCLUSION

The first portion of the research defines Industry 5.0 from academic and business viewpoints. Next, it examines Industry 5.0's technology and then its applications to demonstrate its features. The study explores the challenges of implementing Industry 5.0, which maximises human and machine efficiency. We also discuss where Industry 5.0 is going and which industries need improved management to capitalise on it soon. User interfaces enable "human-machine interaction" between humans and machines. Natural user interfaces like gestures are becoming more popular as machines can be controlled intuitively. Industry 5.0's solution is to put people at the centre of technology and systemic changes.

REFERENCES

Majumdar A, Garg H, Jain R (2021) Managing the barriers of industry 4.0 adoption and implementation in textile and clothing industry: Interpretive structural model and triple helix framework. Comput Ind 125:103372.

Pathak A, Kothari R, Vinoba M, Habibi N, Tyagi VV (2021) Fungal bioleaching of metals from refinery spent catalysts: A critical review of current research, challenges, and future Deleteddirections. J Environ Manage 80:111789.

3 Investigation the Key Enablers of Industry 5.0 Adoption in Manufacturing for Sustainable Development

Rekha Goyat and Mahipal Singh

3.1 INTRODUCTION

In today's world, developments of industries play a major role in the progress and growth of the world. In the 18th century, the development of industry started, but there were a lot of hurdles and barriers in terms of supply chain management, changing environment, lack of technological invention etc. [1]. Initially, production of products like clothing, textiles, weapons, food items, tools, manufacturing and musical instruments was handcrafted by the workers [2]. However, with the increasing demand for products and competition among industries, most organizations focused on adopting mass production using machinery. From the start of the Industrial Revolution Industry 3.0, the robot played a major role in connecting devices that enable human-robot interaction to perform the assigned task. While Industry 4.0 was focused on human-robot collaboration within production lines, Industry 5.0 is about to support human-centric processes that promote resilience, and sustainability between labour and robotic systems. All industry revolutions majorly focused on increasing productivity with higher accuracy but to maintain sustainability with beneficial coexistence and shared prosperity is the main motive of Industry 5.0 [3].

This motivates the teamwork among human labour and tools along with the safety and security of the worker, their skill development and creative potential. The influential growth of businesses can be increased positively with trained professionals and adaptive and safe manufacturing [4]. The innovation of industrialization gives a fresh way of thinking about production that has productive, commercial implications. As a result, organizations that do not adopt Industry 5.0 would quickly become outdated, unable to manage the profitable advantage with business [5]. Also, technologies are increasing day by day which represents the growth of new concepts that are never-ending. As a result, adapting every organization's process and introducing digitization that would ensure that an organization remain feasible. Innovative technology is the way to transform outdated approaches within the industries and to replace that with more creative and computer-controlled technologies. Traditional manufacturing

DOI: 10.1201/9781003535188-3

organizations are totally based on small and medium enterprises that have limited facilities to invest in modern technologies. Although these sectors totally depend on product customization and a small cycle of manufacturing, these sectors are unable to take full benefits of advancement in technology [6]. The way through which companies can shift from traditional to digital manufacturing requires high investment and modern technologies. For small-scale companies, it is very difficult and challenging to employ that much money for revolution. Therefore, researchers and academia have been working on the enablers of Industry 5.0 adoption that mitigate the challenges of Industry 4.0 [7]. One of the common barriers to Industry 4.0 adoption is a lack of knowledge which causes the failure of awareness regarding the future benefits of new technologies. Lack of skill and knowledge of staff members within an organization is also a challenging factor which affects the proper functioning of digitization within the company. To avoid such factors of digital manufacturing implementation, Industry 5.0 is introduced with various enablers to improve the market and global society trends, labour lifestyle and sustainability of the system [8]. As Industry 5.0 is the most emerging area of research, in this chapter we have highlighted possible key enablers of implementation of Industry 5.0 within the organization.

3.2 INDUSTRIAL GROWTH AND DEVELOPMENT

In industry a major change came with the evolution of the First Industrial Revolution (Industry 1.0) in the 18th century. In this revolution, items and products were being produced by machines. In England it was firstly started in 1760 and marked a change from handmade products to machinery products that affect various industries like mining, textile, manufacturing, glass company, agriculture and so on [9]. The first revolution of the industry makes a change in human lifestyle by decreasing the labour need and responsibility by increasing expertise in the working atmosphere. After that second industry called Industry 2.0 revolution came with its own profits from 1871 to 1914 that introduced the concept of faster transfer of innovative ideas and technologies. Industry 2.0 increases economic growth and productivity as machines replace factory labour, but it also increases unemployment [10]. In the 1970s, third revolution termed Industry 3.0 came up with ideas of automation in production processes to reduce human intervention within the factory. Figure 3.1 denotes different industrial paradigms.

The main motive of this revolution is mass production through automation, digital logic and integrated chips to make the whole process of manufacturing automatic [11]. Pre-programmed robots within executing processes were the best example of the Industry 3.0 revolution. Advancement and innovation in technology are transforming the way of traditional production to a fully automatic procedure. In 2011, the fourth revolution (Industry 4.0) was widely presented at the Hannover Fair and introduced the concept of smart industrialization. In this revolution, the processes and industrialization are strongly supported by digitalization and sensing technologies. In Industry 4.0, various sensing devices are deployed within the working industries where data can be collected by the sensor devices and forwarded towards the controlling authority. Industry 4.0 is based on human-machine interaction but somehow compromises with sustainability. Nowadays, Industry 5.0 is the most emerging innovation that

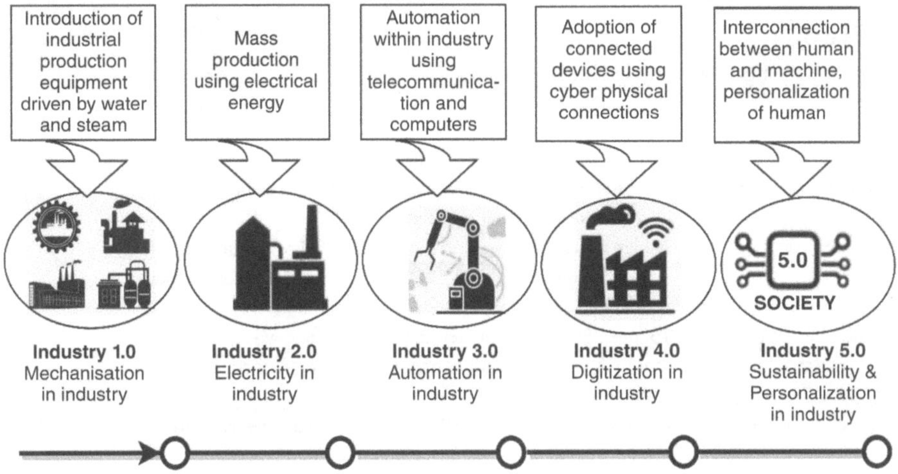

FIGURE 3.1 Industrial Revolution.

represents the humanized version of technological transformation within the industries by considering the present and future perspective of the labour and society and their needs [12]. In this revolution, the new keyword "Sustainability" has been introduced for sustainable growth of industry by optimization of energy consumption, raw materials, processing units and utilization of products. The idea of Industry 5.0 appeared as the extended version of industry 4.0 which focuses on sustainability, human centricity and resilience. Industry 5.0 focused on the skills of labour, their knowledge and their capability to deal with machinery, tools and robots by taking care of flexibilities within the production processes with environmental aspects [13].

3.3 KEY FEATURES OF INDUSTRY 5.0

Industry 5.0 basically works for three important pillars in the area of industrialization. The first pillar is human-centric in which the unique talent and empowerment of each individual labour are considered for industrialization growth. The main motive of Industry 5.0 is to support potential of humans in addition to the machinery and technologies. The second pillar is sustainability within the organization for preserving natural resources, reducing energy wastage and environmental protection [14]. The third pillar is resilience within industries that works seamlessly during crisis time and responds immediately to market changes. Key features of adopting Industry 5.0 within the organization and these features are tabulated in Table 3.1.

3.4 KEY ENABLERS OF INDUSTRY 5.0

Based on the literature, it has been identified that there are various key points which is more crucial for the transition from Industry 4.0 to Industry 5.0. These key points can be used as key enablers for Industry 5.0 implementation. Figure 3.2 denotes key enablers of Industry 5.0.

TABLE 3.1
List of Key Features of Industry 5.0

Key Feature	Description	Reference
Human-centric approach	To prioritize the capability and hidden talent of every individual within the organization and emphasize the collaboration between humans and machines.	[2]
Advanced robotics	Assign tasks collaboratively to humans and robots so that repetitive and dangerous tasks can be done by robots to make a more adaptable and user-friendly environment.	[5]
Artificial intelligence (AI) and machine learning (ML)	To make processes smarter and faster, AI and ML can be used for process optimization, maintenance prediction and more human-beneficial features.	[6]
Internet of Things (IoT)	IoT devices are able to deliver data on a real-time basis allow better monitoring and automation within processes.	[7]
Big data and analytic	Data collected by IoT sensor then analysed to find insight for better innovation and efficiency.	[8, 9]
Augmented reality and virtual reality	To improve the maintenance, training and quality control within the processes augmented reality and virtual reality play a very crucial role.	[11]
Sustainable and resilient manufacturing	Sustainability within the processes to manage the natural resources and reduce overall energy consumption is the main key feature of Industry 5.0	[12]
Customization and personalization	To produce the highly customized and personalized products with respect to quickly adapting the choice of the customer.	[15]
Cybersecurity	Cybersecurity is essential to protect the sensitive data and to maintain the integrity of data.	[16]
Education and training	Industry 5.0 includes upskilling and reskilling of the labour within the companies so that the worker can update themself with new technologies.	[17]

Here is the list of key enablers:

1. *Human-machine collaboration:* It represents the interaction between humans and machinery along with computers, technology and robots to complete the assigned processes. This collaboration can be utilized in various sectors such as education, manufacturing, education and agriculture. Human-machine collaboration would offer complementary skills, mode of interaction, automation within processes, transparency and ethical consideration [18].
2. *Employee empowerment and well-being:* Integration of innovation of humans with technologies offers more personalized and sustainability within the processes. The needs of every individual are achieved with customized and well-being programmes. Industry 5.0 that not only capable of improving sustainability but also improving the labour lifestyle and satisfaction in the workplace [19].

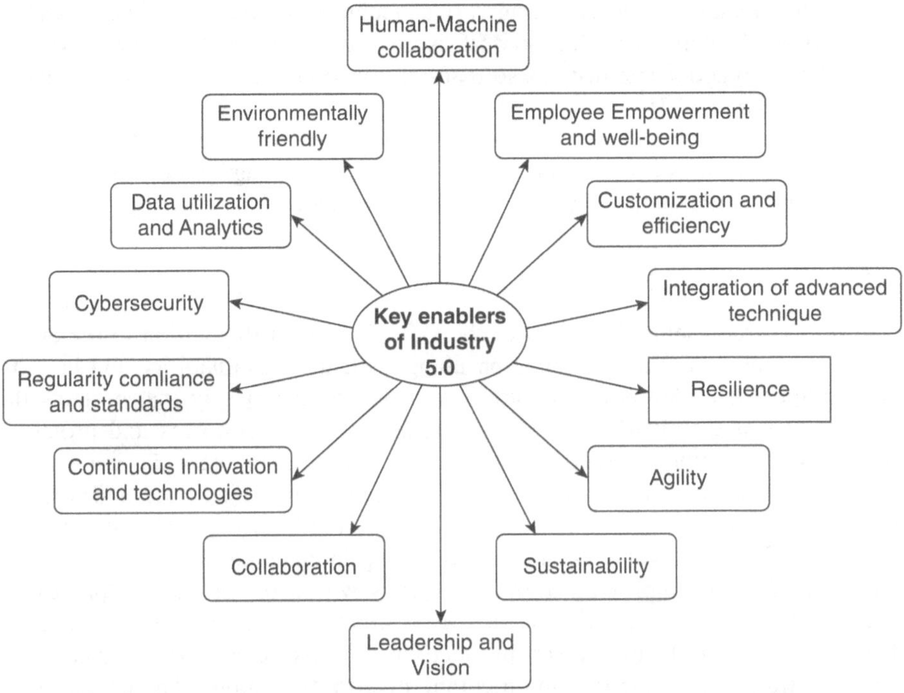

FIGURE 3.2 Key enablers of Industry 5.0.

3. *Integration of advanced technology:* Industry 5.0 highlights the interaction between humans and machines to develop more creative, efficient, well-structured, personalized and sustainable systems within the production. The main motive of this integration increases the operational capabilities of machines and better experience of workers in the workplace [8].
4. *Customization and flexibility:* These two features are the main principle of Industry 5.0 that highlights the capability of each product, service and working conditions should meet required needs and preferences. The incorporation of innovative technologies facilitates a more creative, adaptive and responsive working environment in the industrial landscape [15].
5. *Sustainability and Environmental responsibility:* To improve sustainable growth, Industry 5.0 mainly focuses on minimizing waste and carbon footprints and reducing energy consumption. This approach reflects a commitment to reduce the ecological footprints while encouraging social and economic well-being [17].
6. *Agility and Resilience:* This enabler refers to the capability of industrial processes to predict, be ready with products, respond and able to recover from distractions while maintaining continuity in the operation and giving protection to the workers [20].
7. *Data utilization and analytics:* Properly utilizing and analysis of data is very crucial to smart decisions, improving efficiency and innovative

technologies. By incorporating various technologies like IoT, big data and artificial intelligence, Industry 5.0 is able to transform the collected raw data into actionable insights so that optimization within processes can be achieved [21–23].

8. *Cybersecurity:* Cybersecurity ensures the protection of confidential data against various cyber frauds. As the connection among systems increases, the protection of data and processes becomes paramount [24, 25].

3.5 CONCLUSION

Industry 5.0 represents the initial of the Fifth Industrial Revolution in the era of advanced technology and automation along with the sustainability and human-centric approach. The processes are completed in such a way that protects the environment and reduces energy consumption, carbon footprints and promotes human and machine interaction for industrial growth. Industry 5.0 emphasizes various key enablers such as sustainability, human-centric, resilience, and collaboration that facilitate a transformative shift within the industries. The Industry 5.0 technologies enable seamless interaction among humans and machines for more customization and improving overall efficiency within the production processes. With the integration of human innovation and creativity with precise technologies, Industry 5.0 not only focuses on productivity but also considers environmental sustainability and is able to remain competitive to disruptions. The emphasis on growth, ethical and inclusive development ensures the importance of technological development that leads to more reasonable and advanced industrial systems. This chapter concludes that key enablers of Industry 5.0 enable an innovative approach to industries that coordinates technology capabilities with human value and environment management.

REFERENCES

1. Wang, Baicun, Huiying Zhou, Xingyu Li, Geng Yang, Pai Zheng, Ci Song, Yixiu Yuan, Thorsten Wuest, Huayong Yang, and Lihui Wang. "Human digital twin in the context of Industry 5.0." *Robotics and Computer-Integrated Manufacturing* 85 (2024): 102626.
2. Alves, Joel, Tânia M. Lima, and Pedro D. Gaspar. "Is industry 5.0 a human-centred approach? A systematic review." *Processes* 11, no. 1 (2023): 193.
3. Khan, Moin, Abid Haleem, and Mohd Javaid. "Changes and improvements in Industry 5.0: A strategic approach to overcome the challenges of Industry 4.0." *Green Technologies and Sustainability* 1, no. 2 (2023): 100020.
4. Alojaiman, Bader. "Technological modernizations in the industry 5.0 era: A descriptive analysis and future research directions." *Processes* 11, no. 5 (2023): 1318.
5. Adel, Amr. "The convergence of intelligent tutoring, robotics, and IoT in smart education for the transition from industry 4.0 to 5.0." *Smart Cities* 7, no. 1 (2024): 325–369.
6. Baz, Abdullah, Riaz Ahmed, Suhel Ahmad Khan, and Sudesh Kumar. "Security risk assessment framework for the healthcare industry 5.0." *Sustainability* 15, no. 23 (2023): 16519.
7. Tzampazaki, Maria, Charalampos Zografos, Eleni Vrochidou, and George A. Papakostas. "Machine vision—Moving from Industry 4.0 to Industry 5.0." *Applied Sciences* 14, no. 4 (2024): 1471.

8. Haloui, Doha, Kenza Oufaska, Mustapha Oudani, and Khalid El Yassini. "Bridging Industry 5.0 and Agriculture 5.0: Historical perspectives, opportunities, and future perspectives." *Sustainability* 16, no. 9 (2024): 3507.
9. Ghobakhloo, Morteza, Mohammad Iranmanesh, Masood Fathi, Abderahman Rejeb, Behzad Foroughi, and Davoud Nikbin. "Beyond Industry 4.0: A systematic review of Industry 5.0 technologies and implications for social, environmental and economic sustainability." *Asia-Pacific Journal of Business Administration* (2024). Vol. ahead-of-print No. ahead-of-print. doi: 10.1108/APJBA-08-2023-0384
10. Chidozie, Bernardine, Ana Ramos, José Ferreira, and Luis Pinto Ferreira. "The importance of digital transformation (5.0) in supply chain optimization: An empirical study." *Production Engineering Archives* 30, no. 1 (2024): 127–135.
11. Tallat, Raiha, Ammar Hawbani, Xingfu Wang, Ahmed Al-Dubai, Liang Zhao, Zhi Liu, Geyong Min, Albert Y. Zomaya, and Saeed Hamood Alsamhi. "Navigating industry 5.0: A survey of key enabling technologies, trends, challenges, and opportunities." *IEEE Communications Surveys & Tutorials* 26, no. 2 (2023): 1080–1126. doi: 10.1109/COMST.2023.3329472
12. Fatorachian, Hajar. "Sustainable Supply Chain Management and Industry 5.0." In *Eco-Innovation and Sustainable Development in Industry 5.0*, pp. 172–197. IGI Global, 2024.
13. Nazir, Hassan, Khan Anwar Kamal, Ansar Abbas, Muhammad Raza, Muhammad Yahya, and Muhammad Ibrahim. "Unlocking insights: Structural equation modeling approach to unveil the nexus between industry 5.0 evolution and sustainable supply chain dynamics." *Kurdish Studies* 12, no. 4 (2024): 1095–1107.
14. Sheela, S. Christina, Satyajee Srivastava, S. Vinod Kumar, K. Aravinda, Navdeep Singh, and Akhmadbek Yusupov. "Strategic Implementation of Big Data Analytics for Industry 5.0's Sustainable Expansion." In *Powering Industry 5.0 and Sustainable Development Through Innovation*, pp. 276–291. IGI Global, 2024.
15. Supriya, Y., Dasari Bhulakshmi, Sweta Bhattacharya, Thippa Reddy Gadekallu, Pratik Vyas, Rajesh Kaluri, S. Sumathy, Srinivas Koppu, David J. Brown, and Mufti Mahmud. "Industry 5.0 in smart education: Concepts, applications, challenges, opportunities, and future directions." *IEEE Access* 12 (2024): 81938–81967., doi: 10.1109/ACCESS.2024.3401473
16. Leng, Jiewu, Weinan Sha, Baicun Wang, Pai Zheng, Cunbo Zhuang, Qiang Liu, Thorsten Wuest, Dimitris Mourtzis, and Lihui Wang. "Industry 5.0: Prospect and retrospect." *Journal of Manufacturing Systems* 65 (2022): 279–295.
17. Konstantopoulos, Georgios, Costas A. Charitidis, Miguel A. Bañares, Raquel Portela, Nikolaj Zangenberg, Ennio Capria, Marco Sebastiani et al. "Materials characterisation and software tools as key enablers in Industry 5.0 and wider acceptance of new methods and products." *Materials Today Communications* 36 (2023): 106607.
18. Pizoń, Jakub, and Arkadiusz Gola. "Human–machine relationship—Perspective and future roadmap for industry 5.0 solutions." *Machines* 11, no. 2 (2023): 203.
19. Khaled, Ghazy, and Fedorova Alena. "The evolution of well-being approach within the industry 5.0 concept." *Human Progress* 8, no. 3 (2022): 12.
20. Aheleroff, Shohin, Huiyue Huang, Xun Xu, and Ray Y. Zhong. "Toward sustainability and resilience with Industry 4.0 and Industry 5.0." *Frontiers in Manufacturing Technology* 2 (2022): 951643.
21. Maddikunta, Praveen Kumar Reddy, Quoc-Viet Pham, B. Prabadevi, Natarajan Deepa, Kapal Dev, Thippa Reddy Gadekallu, Rukhsana Ruby, and Madhusanka Liyanage. "Industry 5.0: A survey on enabling technologies and potential applications." *Journal of Industrial Information Integration* 26 (2022): 100257.
22. Czeczot, Grzegorz, Izabela Rojek, Dariusz Mikołajewski, and Belco Sangho. "AI in IIoT management of cybersecurity for industry 4.0 and industry 5.0 purposes." *Electronics* 12, no. 18 (2023): 3800.

23. Raja Santhi, Abirami, and Padmakumar Muthuswamy. "Industry 5.0 or industry 4.0S? Introduction to industry 4.0 and a peek into the prospective industry 5.0 technologies." *International Journal on Interactive Design and Manufacturing (IJIDeM)* 17, no. 2 (2023): 947–979.
24. Yin, Shi, and Yuanyuan Yu. "An adoption-implementation framework of digital green knowledge to improve the performance of digital green innovation practices for industry 5.0." *Journal of Cleaner Production* 363 (2022): 132608.
25. Thomas, Asha, Yash Chawla, Arup Varma, and Przemek Szleter. "The Role of Business Leaders in Industry 5.0." In *The International Research & Innovation Forum*, pp. 629–640. Springer International Publishing, 2023.

4 Challenges to Industry 5.0 and Removal Actions

Upasna Goyal, Geeta Sachdeva, and Rajeev Rathi

4.1 INTRODUCTION

The landscape of industrial and organizational operations is undergoing a rapid transformation, thanks to the integration of advanced technologies that have fundamentally reshaped business operations. In recent years, a significant shift has occurred from "Industry 4.0" to "Industry 5.0." According to Forbes (2022) for many companies, Industry 4.0 is either the forthcoming priority or the trend with which they are currently aligning their strategies. This phase, known as the Fourth Industrial Revolution (4IR), is driven by advancements in information technology. Core components of Industry 4.0 include automation, robotics, big data analytics, intelligent systems, virtualization, artificial intelligence (AI), machine learning (ML), and the Internet of Things (IoT). The transition to Industry 5.0, with its potential to transform the manufacturing industry, offers a bright and promising future for industrial operations (Ghobakhloo et al., 2024).

The transition from the 4IR to the Fifth Industrial Revolution (5IR) marks a significant change in how we approach industrial processes. Industry 5.0 is about harnessing the collaborative potential between humans and machines and transforming the manufacturing sector into a more adaptable, sustainable, and human-centric industry (Leng et al., 2022). This evolution fosters economic growth while positively impacting society and the environment. By reemphasizing the partnership between humans and machines, Industry 5.0 promises to unlock extraordinary levels of innovation and productivity in the industrial world, offering a brighter future for businesses and workers alike (Alpaslan Demir and Cicibaş, 2019).

However, as digital transformation continues to permeate workplaces, it also introduces challenges that require our attention. One such challenge is the pervasive issue of "cyberloafing", which has emerged as a significant problem. Its potential to undermine productivity, employee morale, and organizational efficiency cannot be overstated (Tandon et al., 2022). As the boundaries between professional and personal life blur, particularly with the rise of remote and hybrid work models, cyberloafing has become a pressing concern (Chakraborty et al., 2024). The importance of understanding the relationship between cyberloafing and the latest industrial revolutions cannot be overstated (Shaddiq et al., 2021). It is not just an academic exercise but a crucial step in developing effective strategies and policies. This understanding can help us harness the advantages of Industry 5.0 while mitigating the adverse effects of cyberloafing, thereby ensuring a smooth transition to the next phase of the industrial revolution (Gajdzik, 2023).

DOI: 10.1201/9781003535188-4

In this dedicated chapter, the authors delve into the history of the Industrial Revolution, tracing its evolution through various phases, highlighting key milestones, and emphasizing the distinction between 4IR and the emerging 5IR. Furthermore, it explores the multifaceted nature of Industry 5.0. This includes its attention to sustainability, resilience, and a more personalized approach to production. A significant focal point of this chapter emerges around the challenges posed by the latest industrial phase, prominently featuring the phenomenon of cyberloafing. With the increased integration of digital technologies in this phase, there is potential for this issue to be exacerbated.

4.2 LITERATURE REVIEW

The first industrial revolution started in Britain at the end of the 18th century and was initiated by the introduction of James Watt's steam engine. This era marked a significant transition from handmade goods to machine-driven manufacturing. Notable advancements included the steam engine, mechanized textile production, and iron and steel production improvements. The revolution resulted in the growth of factories and urban areas as individuals moved from rural areas to cities searching for jobs. Although it stimulated substantial economic growth and technological advancement, it also resulted in poor working conditions and environmental harm. These industrial methods and technologies gradually spread to other regions of Europe and the United States, laying the foundation for modern industrial society. The Second Industrial Revolution, occurring between 1870 and 1914, marked a period of significant technological advancements that transformed various industries and significantly improved human welfare. Unlike its predecessor, this era saw a surge in real wages, improved living standards, and a shift in technological leadership to a more dispersed geographical focus. The rapid technological progress was driven by the accumulation of valuable knowledge, leading to widespread innovations that impacted the daily lives of both the middle and working classes. The production organization also changed with the rise of large corporations benefitting from economies of scale. Communication was also revolutionized by the invention of the telegraph and telephone, allowing for instant long-distance communication and further integrating global economies and societies. The third industrial revolution, often called the Digital Revolution, commenced in the 1950s and marked a significant shift in manufacturing towards digitization, customization, and automation. This era is characterized by a transition from mechanical and analogue technologies to digital electronics. Unlike the mass production focus of the past, the future factory will emphasize mass customization, producing smaller batches tailored to individual customer preferences. This transformation is expected to bring about global development, create wealth, and spread economic growth worldwide through innovations in economic development, transportation, etc. Additionally, the Third Industrial Revolution is part of the broader information age, which began in the mid-20th century and centres on information technology. Consequently, Industry 5.0 not only benefits companies but also their employees. According to the literature review, 5IR will be marked by a revitalized and broadened sense of purpose, extending beyond merely generating profit through producing goods and services.

Instead of beginning with the latest technology and exploring how it can improve efficiency, a human-centred approach in industry prioritizes fundamental human needs and interests in the production process. This perspective shifts the question from what we can achieve with new technology to how technology can benefit us. Instead of requiring workers to adjust their skills to keep up with fast-changing technology, we aim to use technology to tailor the production process to worker's needs. This goal ensures that technological integration supports and adds human experience in the workplace, safeguarding fundamental rights such as privacy and dignity.

To align with environmental limits, the industry must be sustainable. Sustainability entails establishing circular processes that reuse and recycle natural resources, minimizing waste and environmental impact. It also involves reducing energy consumption and greenhouse gas emissions to prevent resource depletion and environmental degradation. It's a crucial responsibility to ensure that we meet our current needs without jeopardizing the ability of future generations to meet theirs. Technologies such as AI and additive manufacturing can significantly contribute to this cause by enhancing resource efficiency and reducing waste.

As businesses and organizations worldwide are still in the process of understanding and implementing various aspects of Industry 4.0, the rapid emergence of Industry 5.0 has caused some confusion. This new term appeared several years ago, describing an updated version of the Industrial Revolution that responded to the experiences and lessons learned from implementing 4IR solutions. According to the global tech and business community promoting this concept, Industry 5.0 should be viewed as an enhancement, addition, and potentially modification of Industry 4.0, with some even suggesting it could be considered version 4.1. Here are some key points highlighting the differences between Industry 4.0 and 5.0.

Industry 4.0 focuses on improving manufacturing and production efficiency through automation and advanced technologies. This shift aims to develop interconnected and intelligent production systems that streamline operations. Transitioning to Industry 5.0, the focus expands to encompass the creation of sustainable and environment-friendly manufacturing processes, ensuring that technological advancements are made to respect and prioritize environmental sustainability. Industry 4.0 necessitates IoT, AI, and ML competencies for automating tasks and decision-making. As we move to Industry 5.0, these cutting-edge technologies are integrated with human skills and creativity, acknowledging the importance of human intuition and innovative thinking in driving production forward.

Digital twins and simulation tools play crucial roles in Industry 4.0 by enhancing production processes through detailed virtual representations. Industry 5.0 further broadens the technological landscape by incorporating advanced technology like nanotechnology and biotechnology. These innovations facilitate the development of new materials and products. In the era of Industry 4.0, robots and autonomous machines are widely employed for tasks that are repetitive, hazardous, or require precise execution, thereby improving safety and productivity. Transitioning to Industry

5.0, there is an emphasis on nurturing new skills and capabilities among human workers, enabling them to work efficiently and effectively with advanced technologies and embrace more creative and complex roles. Building upon the data-centric approach of Industry 4.0, which focuses on using data and analytics to optimize processes, Industry 5.0 advances this concept by emphasizing the significance of human interaction and collaboration. Although data remains vital, the new focus is on harnessing human creativity and problem-solving skills with technological tools.

The critical difference between the Industry 5.0 concept and Industry 4.0 is its emphasis on integrating the "human touch" into machines and leveraging new technologies to empower human workers better. Here are some notable comments from leading voices on the same.

1. *Frost and Sullivan's Vision:* Frost and Sullivan predict that Industry 5.0 will mark a significant shift towards restoring the role of human workers in manufacturing. Unlike Industry 4.0's fully automated focus, this new paradigm emphasizes a more balanced integration of human expertise alongside technological advancements on the shop floor (Zutshi, 2019).
2. *Elon Musk's Viewpoint:* Elon Musk (CEO of Tesla) has criticized the Industry 4.0 ideology for its heavy focus on automation and robotics in manufacturing. He tweeted that Tesla's attempt to rely extensively on automated processes was a mistake (Musk, 2018), particularly for the Model 3 sedan production, which led to unforeseen challenges and inefficiencies, a period he famously referred to as "production hell" (CBS News, 2018). Musk emphasized that relying too much on automation slowed down rather than accelerated manufacturing processes. His experience highlights the importance of finding a balance between technological advancements and the capabilities of human workers in industry.
3. These ideas reflect both Frost and Sullivan's and Elon Musk's views on the shift towards Industry 5.0, which stresses a reevaluation of automation and a renewed focus on human-centred approaches in manufacturing.

4.3 CHALLENGES TO ADOPT INDUSTRY 5.0

Industry 4.0, introduced by Klaus Schwab of the World Economic Forum in 2016, sought to minimize human involvement in manufacturing through increased automation. However, starting in 2020, the focus has shifted to Industry 5.0, which emphasizes collaboration between humans and machines. Each of these industrial transformations brings its own set of challenges. The difficulties of integrating advanced systems, maintaining cybersecurity, and managing the human-machine interface are some significant challenges manufacturers encounter when implementing these transformations. To address these challenges and facilitate this shift, Info-Tech Research Group has published fresh perspectives in its most recent industry blueprint, "Transform Manufacturing by Mitigating the Challenges of Industry 4.0 and 5.0." Details are illustrated in Figure 4.1.

Shukla (2024), principal research director at Info-Tech Research Group, emphasized, "The global manufacturing landscape is rapidly evolving, and the integration

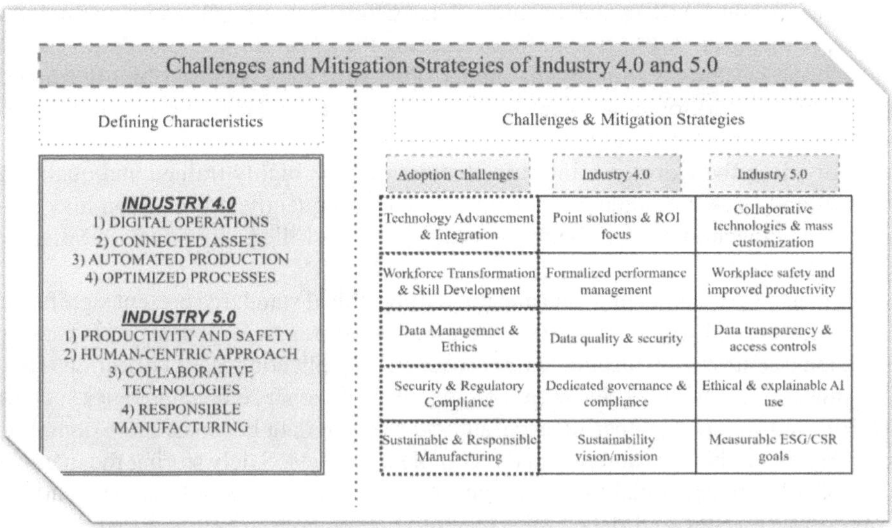

FIGURE 4.1 Challenges in the adoption of Industry 4.0 and Industry 5.0 technologies.

of Industry 4.0 technologies has resulted in a more interconnected and data-driven manufacturing environment," and added, "Furthermore, the leap to Industry 5.0 introduces additional layers of complexity, emphasizing customization, efficiency, and more significant collaboration between humans and machines."

The new resource (Figure 4.1) offers essential guidance and tactics for tackling these issues. It features case studies and expert insights illustrating how top companies successfully address these challenges and foster a culture of innovation. Info-Tech identifies six fundamental obstacles that hinder manufacturers from embracing Industry 4.0 and 5.0 technologies.

Adopting new technology in the manufacturing industry is a gradual and demanding process involving several advanced technologies collectively known as Industry 5.0, such as tailored software linking factories, real-time data analytics, collaborative robots, 3D printing, AI, the IoT, and cloud computing. These technologies require significant investment, as acquiring equipment like UR Cobots (collaborative robots produced by Universal Robots (UR) company) and training employees for new roles can be costly. Upgrading production lines to Industry 5.0 standards can be challenging, even for well-funded companies, due to the rapid pace of technological change. Consequently, businesses that cannot afford these upgrades or adapt quickly may fall behind. Additionally, obtaining high-quality, reliable data and integrating various data repositories present significant challenges in manufacturing systems (Thoben et al., 2017).

The evolving workforce landscape, particularly with the integration of robots and AI, requires employees to adapt by acquiring new skills. As robots and intelligent machines become more prevalent, employees must learn to collaborate effectively with these technologies, including gaining technical expertise and honing soft skills. The emergence of new roles, such as programming and managing industrial robots, will create positions like Chief Robotics Officer or Chief AI Officer (Cojocariu, 2019), crucial for overseeing the integration, and operation of AI and robotic systems within organizations. This shift necessitates a highly trained and qualified workforce in these new areas of expertise. However, this trend also exacerbates work polarization by reducing the number of middle-skill jobs, increasingly dividing the workforce into two distinct groups: highly trained and qualified workers and low-paid, unqualified employees. Consequently, this division may narrow the socioeconomic gap between skilled and unskilled individuals (Dattatray Zhodage, 2022).

In Industry 5.0, managing data and upholding ethical standards present significant challenges crucial for effective data use and ethical practices. Industry 5.0 focuses on human-centric and sustainable approaches, complicating data collection, storage, processing, and analysis. With the integration of advanced technologies such as AI, IoT, and robotics, gathering accurate and relevant data becomes more complex, often leading to biased sampling and incomplete datasets. Safely storing the massive amounts of data generated by these technologies demands state-of-the-art solutions to prevent breaches and data loss, especially as cybersecurity threats grow more sophisticated and pervasive. Analysing this data to extract valuable insights requires advanced analytical tools and methodologies to ensure high data quality. Ethical considerations, including informed consent, privacy protection, and transparency, are vital for maintaining trust and regulatory compliance. Addressing cybersecurity risks is essential to protect sensitive information from cyberattacks and unauthorized access, ensuring data integrity and security throughout its lifecycle. Overcoming these challenges is crucial for harnessing data-driven insights and making informed, ethical decisions in Industry 5.0.

Making eco-friendly manufacturing a reality in the age of advanced technology isn't easy. The 5IR aims to unite people and machines like AI and internet-connected devices to create a more efficient and sustainable industry. However, integrating these new technologies into existing factories is often complicated and expensive, especially for small and medium-sized enterprises (SMEs). Fixing problems like resource waste and high energy use requires a lot of money upfront and training workers on how to use the new systems. Additionally, ensuring the latest technologies work seamlessly with current systems adds another layer of complexity. Despite all these challenges, switching to sustainable practices is crucial to reduce our environmental impact and ensure the manufacturing industry stays strong in the long run.

The widespread adoption of Industry 5.0 technologies presents significant challenges for industrial sector security and regulatory compliance. The cyber threat landscape expands dramatically as industries integrate advanced technologies and interconnect a broader range of assets. This necessitates robust safeguards for industrial control systems and intellectual property protection to mitigate the risk of breaches. The evolving regulatory environment across various global markets further heightens the complexity. Stringent compliance measures are essential to minimize legal and operational risks associated with non-adherence. In this context, developing and implementing robust cybersecurity frameworks is paramount.

TABLE 4.1
Barriers to Industry 5.0

Barriers	Description	References
Technological Integration	Integrating different technologies poses a significant challenge, emphasizing the importance of ensuring seamless communication and interoperability among these systems.	George and Hovan George (2023)
Skill Gap	There is an insufficiency of skilled professionals to manage advanced technologies adeptly.	Adel (2022)
Cybersecurity Risks	Heightened susceptibility to cyber-attacks due to increasingly interconnected advanced systems.	Hassan et al. (2024)
Job Displacement	As automation continues to advance there is a concern about the potential impact on job security.	George and Hovan George (2023)
Expensive	Investing in new technologies often requires a substantial financial commitment.	Hassan et al. (2024)
Data Privacy Concerns	Managing and safeguarding large volumes of sensitive data comes with significant risks.	Tallat et al. (2024)
Environmental Value Measurement	Establishing a universal standard for measuring environmental metrics faces considerable challenges.	Rajesh (2023)

The primary obstacle to implementing digital-first initiatives within organizations is overcoming employee resistance and adaptation hurdles. This resistance can be attributed to many factors, including apprehensions regarding job displacement, a lack of familiarity with novel technologies, and a profoundly ingrained organizational culture that prioritizes established methodologies. These impediments hinder the seamless integration and execution of digital-first strategies, potentially delaying the organization's ability to capitalize on technological advancements for a competitive edge and enhanced operational efficiency. Addressing these challenges necessitates a nuanced comprehension of cultural and organizational dynamics and a strategic plan to mitigate resistance and foster an environment conducive to digital transformation. Table 4.1 denotes barriers to Industry 5.0.

The Internet has significantly transformed how people live, work, and learn. It has introduced convenience and revolutionized communication. In professional settings, it has also opened up new opportunities for employees to participate in activities that are unrelated to work, and occasionally, these activities can be non-productive. Blanchard and Henle (2008) observed a growing trend of employees spending more time on non-work-related activities online. These activities encompass a wide range of online behaviour, such as "shopping, blogging, gaming, instant messaging, gambling, booking vacations, viewing pornography, checking and sending private emails, personal investing, participating in auctions, hunting for jobs, and visiting websites related to sports, news, entertainment, as well as general web surfing" (Vitak et al., 2011). In general, the term "cyberloafing" is used in the literature to refer to behaviours where an employee participates in electronically mediated activities that their immediate supervisor would not deem as related to their job (Askew et al., 2014). Various terminologies are employed in academic studies to depict this phenomenon, like cyberslacking, cyberbludging, goldbricking, internet deviance, junk computing, and online loafing (Kim and Byrne, 2011).

However, academia began to acknowledge the issues arising from employees' improper internet usage and exploitation shortly after Lim (2002) defined the concept of cyberloafing. Since then, research in this domain has garnered attention (Tandon et al., 2022). Studies on cyberloafing have been carried out globally, with contributions from scholars in the United States, China, Europe, and the Middle East, indicating that it is a widespread phenomenon in workplaces worldwide (Wu et al., 2021). Although there is still much to discover about cyberloafing, significant progress has been made over the past two decades (Tandon et al., 2022). Despite the differences in terminology across studies, a common underlying theme is that cyberloafing involves engaging in online activities during periods when one is supposed to be working (Lim and Teo, 2024). Cyberloafing poses a considerable challenge in adopting Industry 5.0, a phase that relies on the collaboration between humans and advanced technologies to create more personalized and efficient manufacturing processes. This integration demands high digital engagement and responsibility from our valued employees. The threat of cyberloafing, which can reduce productivity, increase security risks, and undermine the sophisticated human-machine synergy, is a clear call for responsible digital use.

4.4 BARRIER'S REMOVAL PLANS

Table 4.2 denotes different barrier removal plans.

TABLE 4.2
Barrier Removal Plans

Barriers	Action Plan
Technological Integration	• Develop and adopt open standards and protocols. • Foster public-private partnerships for technology integration.
Skill Gap	• Create continuous learning and upskilling programmes. • Partner with educational institutions for tailored training modules.
Cybersecurity Risks	• Implement robust cybersecurity protocols and continuous monitoring. • Conduct regular security audits and employee training.
Job Displacement	• Develop reskilling and job transition programmes. • Promote the creation of new job roles in the technology and service sectors.
Expensive Implementation	• Offer financial support and subsidies for SMEs. • Encourage shared resource models and collaborative investment.
Data Privacy Concerns	• Establish strict data privacy policies and compliance with GDPR (General Data Protection Regulation). • Implement secure data handling practices.
Environmental Sustainability	• Adopt sustainable practices and green technologies. • Implement circular economy principles in manufacturing and logistics.

4.5 CONCLUSION

5IR marks the subsequent phase of industrial evolution, highlighting the synergy between humans and technology. Unlike 4IR, which concentrates on automation and the IoT, Industry 5.0 emphasizes combining human intelligence with cognitive computing. This evolution aims to foster more personalized, sustainable, and resilient manufacturing processes. Cyberloafing has also evolved with each industrial advance. During the period of Industry 4.0, the digital nature of work environments made it easier for employees to engage in cyberloafing, potentially impacting productivity. In the context of Industry 5.0, where the collaboration between humans and machines is crucial, the implications of cyberloafing may change.

REFERENCES

Adel, A. (2022), "Future of Industry 5.0 in society: Human-centric solutions, challenges, and prospective research areas", Journal of Cloud Computing, Vol. 11 No. 1, 40.

Alpaslan Demir, K. and Cicibaş, H. (2019), The Next Industrial Revolution: Industry 5.0 and Discussions on Industry 4.0, *4th International Management Information Systems Conference Industry,* Vol. 4, pp. 17–20.

Askew, K., Buckner, J.E., Taing, M.U., Ilie, A., Bauer, J.A. and Coovert, M.D. (2014), "Explaining cyberloafing: The role of the theory of planned behavior", Computers in Human Behavior, Vol. 36, pp. 510–519. DOI: 10.1016/j.chb.2014.04.006.

Blanchard, A.L. and Henle, C.A. (2008), "Correlates of different forms of cyberloafing: The role of norms and external locus of control", Computers in Human Behavior, Vol. 24 No. 3, pp. 1067–1084. DOI: 10.1016/j.chb.2007.03.008.

CBS News. (2018), "Tesla CEO Elon Musk, Stressed but 'Optimistic,' Predicts Big Increase in Model 3 Production", *CBS News*, 13 April, available at: https://www.cbsnews.com/news/elon-musk-tesla-model-3-problems-interview-today-2018-04-13/ (accessed 14 June 2024).

Chakraborty, T., Dhir, S., Al-Busaidi, A.S. and Dwivedi, Y.K. (2024), "Can Remote Work Be Adopted When Phubbing and Cyber Loafing Behavior Are on the Rise?", Journal of Global Information Management, IGI Global, Vol. 32 No. 1. DOI: 10.4018/JGIM.341800.

Cojocariu, O. (2019), "Industry 5.0 Opportunities and Challenges: Bring Your Factory into the Future", *Digitalya*, available at: https://digitalya.co/blog/industry-5-opportunities-and-challenges/#:~:text=The%20challenges%20of%20Industry%205.0&text=People%20may%20need%20to%20develop,will%20also%20be%20an%20issue (accessed 13 June 2024).

Dattatray Zhodage. (2022), "Industry 5.0: Introduction, Opportunities and Challenges", *LinkedIn*, 6 November.

Gajdzik, B. (2023), "Industry 5.0 as the upgrade of Industry 4.0: Towards one common concept of industrial transformation", Scientific Papers of Silesian University of Technology. Organization and Management Series, Vol. 2023 No. 181, pp. 131–150. DOI: 10.29119/1641-3466.2023.181.9.

George, A.S. and Hovan George, A.S. (2023), "Partners Universal International Innovation Journal (PUIIJ) Revolutionizing Manufacturing: Exploring the Promises and Challenges of Industry 5.0". DOI: 10.5281/zenodo.7852124.

Ghobakhloo, M., Iranmanesh, M., Fathi, M., Rejeb, A., Foroughi, B. and Nikbin, D. (2024), "Beyond Industry 4.0: A systematic review of Industry 5.0 technologies and implications for social, environmental and economic sustainability", Asia-Pacific Journal of Business Administration. DOI: 10.1108/APJBA-08-2023-0384.

Hassan, M.A., Zardari, S., Farooq, M.U., Alansari, M.M. and Nagro, S.A. (2024), "Systematic analysis of risks in Industry 5.0 architecture", Applied Sciences (Switzerland). DOI: 10.3390/app14041466.

Kim, S.J. and Byrne, S. (2011), "Conceptualizing personal web usage in work contexts: A preliminary framework", Computers in Human Behavior, Vol. 27 No. 6, pp. 2271–2283. DOI: 10.1016/j.chb.2011.07.006.

Leng, J., Sha, W., Wang, B., Zheng, P., Zhuang, C., Liu, Q. and Wuest, T., et al. (2022), "Industry 5.0: Prospect and retrospect", Journal of Manufacturing Systems, Vol. 65, pp. 279–295. DOI: 10.1016/j.jmsy.2022.09.017.

Lim, V.K.G. (2002), "The IT way of loafing on the job: Cyberloafing, neutralizing and organizational justice", Journal of Organizational Behavior, Vol. 23 No. 5, pp. 675–694. DOI: 10.1002/job.161.

Lim, V.K.G. and Teo, T.S.H. (2024), "Cyberloafing: A review and research agenda", Applied Psychology, Vol. 73 No. 1, pp. 441–484.

Musk, E. (2018), "Yes, Excessive Automation at Tesla Was a Mistake. To Be Precise, My Mistake. Humans Are Underrated", *Twitter*, 14 April, available at: https://x.com/elonmusk/status/984882630947753984?lang=en (accessed 14 June 2024).

Rajesh, R. (2023), "Industry 5.0: Analyzing the challenges in implementation using grey influence analysis", Journal of Enterprise Information Management, Vol. 36 No. 5, pp. 1349–1371. DOI: 10.1108/JEIM-03-2023-0121.

Shaddiq, S., Haryono, S., Muafi, M. and Isfianadewi, D. (2021), "Antecedents and consequences of cyberloafing in service provider industries: Industrial Revolution 4.0 and Society 5.0", Journal of Asian Finance, Vol. 8 No. 1, pp. 157–167. DOI: 10.13106/jafeb.2021.vol8.no1.157.

Shukla, S. (2024), "Transform Manufacturing by Mitigating the Challenges of Industry 4.0 and 5.0", *Info-Tech Research Group*, 18 June.

Tallat, R., Hawbani, A., Wang, X., Al-Dubai, A., Zhao, L., Liu, Z. and Min, G., et al. (2024), "Navigating Industry 5.0: A survey of key enabling technologies, trends, challenges, and opportunities", IEEE Communications Surveys and Tutorials, Vol. 26 No. 2, pp. 1080–1126. DOI: 10.1109/COMST.2023.3329472.

Tandon, A., Kaur, P., Ruparel, N., Islam, J.U. and Dhir, A. (2022), "Cyberloafing and cyberslacking in the workplace: Systematic literature review of past achievements and future promises", Internet Research, Vol. 32 No. 1, pp. 55–89. DOI: 10.1108/INTR-06-2020-0332.

Thoben, K.D., Wiesner, S.A. and Wuest, T. (2017), "'Industry 4.0' and smart manufacturing a review of research issues and application examples", International Journal of Automation Technology. DOI: 10.20965/ijat.2017.p0004.

Vitak, J., Crouse, J. and Larose, R. (2011), "Personal internet use at work: Understanding cyberslacking", Computers in Human Behavior, Vol. 27, pp. 1751–1759. DOI: 10.1016/j.chb.2011.03.002.

Wu, J.N., Song, M., Ugrin, J.C., Liu, L. and Zhu, T. (2021), "Cyberloafing research 1997-2019: A citation-based literature review", Organizacija, Vol. 54 No. 2, pp. 98–111..

Zutshi, A. (2019), "Industry 5.0—Bringing Empowered Humans Back to the Shop Floor", *Frost & Sullivan*, 26 November, available at: https://www.frost.com/frost-perspectives/industry-5-0-bringing-empowered-humans-back-to-the-shop-floor/ (accessed 14 June 2024).

5 Industry 5.0 in the Educational System

Improving the Learning Process

Khadija Echefaj, Abdelkabir Charkaoui, and Anass Cherrafi

5.1 INTRODUCTION

Industry 5.0 (I5.0) has arisen as an extension of industry 4.0 with an emphasis on human interest, resilience and sustainability (Nahavandi, 2019). This shift arises from the shortcoming of its predecessor that stresses the implementation of advanced technologies into the manufacturing domain (Breque et al., 2021; Ivanov, 2023). I5.0 sees the presence of human creativity and innovation complementary to the role of machines in enhancing productivity and efficiency (Leng et al., 2022).

The implications of I5.0 go beyond the production domain to various fields. Education is one of the domains that was influenced by this industrial revolution (Gürdür Broo et al., 2022; Kumar et al., 2024). The educational system cannot ignore the environmental changes and continue with the traditional teaching methods. This is because rapid technological advancements are reshaping diverse factors such as lifestyle, markets and jobs (Cherrafi et al., 2022; Danaher, 2017; Echefaj et al., 2024). Furthermore, the global concerns regarding sustainability drive the transformation of institutional practices and pedagogical curricula to form an ecological conscious generation (Kaswan et al., 2024; Lozano et al., 2017). Additionally, I5.0 and education converge in their shared focus on human interest as they place humans at the centre of pedagogical, technological and social development (Logunova and Utkina, 2019; Papetti et al., 2020).

The interest towards I5.0 implications for the educational system has gained a considerable attention recently. It is critical to explore the transition towards the educational system 5.0. This study investigates the integration of I5.0 principles into education. It explores the transformation towards a human-centric, sustainable and resilient educational system. Moreover, the changes brought by technologies 5.0 on the learning process and the challenges are highlighted. A conceptual framework for educational system 5.0 is proposed.

5.2 THEORETICAL BACKGROUND

I5.0 merges the innovative potential of humans with intelligent machines to improve efficiency and productivity (Nahavandi, 2019). Leng et al. (2022) state that I5.0 emphasizes the interactions between cobots and workers in a smart factory. It focuses on

DOI: 10.1201/9781003535188-5

sustainability, resilience and human centricity (Breque et al., 2021). Sustainability stresses the importance of economic growth simultaneously with environment prevention and social equity (Rajeev et al., 2017). The purpose is to fulfil the current and the future generations' needs. Resilience refers to the ability to endure and bounce back from disturbance (Ali et al., 2021). In the lens of I5.0 structures should be resilient in the face of unknown risks (Ivanov, 2023). Human centricity puts the human interest in the centre of focus of industrial processes (Villar et al., 2023). This is because efficiency cannot be achieved without human creativity and innovation. This approach accentuates the need for human-machine collaboration, empowerment, upskilling and user-centred design (Alves et al., 2023; Murphy et al., 2023; Wan and Leirmo, 2023; Xu et al., 2021).

I5.0 technologies incorporate information and operational tools that drive the sustainable and human-centric methods of value creation (Maddikunta et al., 2022). The most popular are artificial intelligence (AI), cognitive AI, Internet of Things (IoT), additive manufacturing (AM), sensors, robots, virtual/augmented reality (V/AR), cloud computing (CC), cyber security (CS), cyber-physical systems (CPS) and human interaction and recognition technologies (HIRT) (Ghobakhloo et al., 2023; Müller, 2020).

The educational system is an important element for the development of each member of the society, communities and nations. It is a complex and dynamic system that prepares students for the future. Curriculum and teaching methods are one of the vital components of the educational landscape (Ebiere Dorgu, 2015). These are in continuous changes from traditional lecture/instruction methods towards more student-centred approaches (Danko and Duarte, 2009). This evolution is explained by the students' need to cultivate more skills and creativity to fit market jobs the requirements (Alghafri and Ismail, 2014).

Furthermore, equity and access are two important elements in the educational system. Several studies have outlined the disparities and differences in education quality and availability between areas (McCowan, 2016; Motala, 2015). In the face of this situation, diverse actions were adopted to minimize this deviation meeting the sustainable development goals. According to Gürdür Broo et al. (2022), technological advancements have made significant changes to the educational system, improving methods and tools.

The integration of I5.0 principles into education has transformative effects on the learning process (Al-Emran and Al-Sharafi, 2022). The topic is gaining more importance in literature as both fields stress the human-centric approach. Table 5.1 shows the differences between I5.0 and Industry 4.0 views regarding education, and Table 5.2 summarizes the findings of the previous works connecting I5.0 and education.

TABLE 5.1
The Shift from Industry 4.0 to I5.0 in Education

Axe	Industry 4.0	Industry 5.0
Focus	Efficiency and productivity	Sustainability, resilience and human centricity
Partnership with the industrial domain	Align teaching content with market workforce needs	Co-development of curriculum and research projects
Learning environment	Traditional classrooms	Flexible and adaptable spaces
Human and machine relationship	Automating learning tasks and replacing the educational staff	Educators and students develop competencies while working with machines
Learning experience	Standardized learning experience	Personalized learning experience

TABLE 5.2
Previous Studies on Education and I5.0

Reference	Focus	Main Findings
Mohamed Hashim et al. (2024)	Investigate the potential of I5.0 in higher education	The adoption of I5.0 is in its infancy stages in university. The implementation of I5.0 requires important investments and extensive research. I5.0 develops research innovation hubs. I5.0 eases the access to technology.
Roy and Ateeq (2024)	Investigate the opportunities and challenges of I5.0 to universities	I5.0 enhances engagement and motivation Social belonging and wellbeing are improved promotion of acquisition and new skills University should implement a digital transformation strategy, build industry partnerships, adopt robust IT structures and develop an innovation culture.
Adel (2024)	Examine the key elements of smart education	The smart education plays an important role in enhancing upskilling, adaptability, self-learning and resilience. The challenges incorporate data privacy, actors' readiness, AI-system biases and the digital divide.
Supriya et al. (2024)	Investigate the role of I5.0 in smart education	I5.0 promotes distance and further learning and engineering education. Investment, privacy, infrastructure and adaptability are challenging.
Ciolacu et al. (2023)	Present the main teaching approaches for skills development	Future engineers should develop curiosity, agile thinking, creativity and problem-solving abilities
Bakkar and Kaul (2023)	Address the link between I5.0 and education 5.0 regarding future skills.	Various players should participate to enhance the educational system. Future skills incorporate: complex problem-solving, critical thinking, creativity, emotional intelligence, adaptability, digital skills, cognitive flexibility, negotiation, …
Bakir and Dahlan (2022)	Explore the responses of higher education institutions in post Covid-19	The environmental changes influence the curriculum design and the leaders' decisions. Leaders are perspective, agile, embracing culture change and able to learn from experience. Leaders are developing more flexible and student-centred curricula.
Gürdür Broo et al. (2022)	Investigate the education engineering in the era of I5.0	Upskilling is needed to deal with I5.0 challenges Four strategies are identified: • Lifelong learning and transdisciplinary education • Sustainability, resilience and human-centric design modules • Hands-on data fluency and management courses • Human-agent/machine/robot/computer interaction experiences
Al-Emran and Al-Sharafi (2022)	Provide an overview of I5.0 in education and highlight implementation challenges	Application of I5.0 in education: smart classroom, adaptive learning software, big data analytics, predictive analytics, wearable technologies and AI-based intelligent scheduling systems. Challenges of I5.0 for education: difficulty in predicting future jobs, investment, access to digital resources, evaluation methods.

Table 5.2 shows that the research on the implications of I5.0 in the education system is relatively new and emergent. It reveals many limitations that require further investigation. The reviewed studies mainly focus on higher education and the challenges that hinder the transition towards education 5.0. They tend to focus on the technological advancement brought by I5.0. However, more discussions regarding the effect of I5.0 principles on education are required. Furthermore, there is a need to develop a conceptual framework that incorporates education 5.0 principles, tools, and challenging aspects that should be considered while evolving towards the new system. Given the nascent nature of the field (Mohamed Hashim et al., 2024), explorative qualitative research is recommended to capture practical insights for a more nuanced understanding.

5.3 METHODOLOGY

In this study, a qualitative approach is adopted to explore the integration of I5.0 into the educational system. As research in this topic is still in its infancy stage and few works are performed to analyse the interactions between I5.0 and education, explorative studies are beneficial. Yin (2009) states that they allow to capture of new insights and initiate the underdeveloped research areas. The findings constitute a first step in investigating the interactions between education and I5.0. Figure 5.1 describes the research protocol.

In the first step, 12 experts are selected based on several criteria to ensure the reliability the validity of the study. This sample size meets the literature threshold as many studies suggested the number of 12 as the minimum to adopt (Braun and Clarke, 2016; Luz Tortorella et al., 2022). The selection criteria include (1) a deep understanding and expertise in the intersection between industry and education, (2) diversity of perspectives, (3) various educational contexts and (4) geographic diversity. First, we have involved educational professionals who play strategic and practical roles in the field. Educators were also involved as they are the first concerned. To ensure that they are suitable, we have chosen educators that are at the same time researchers in the industrial domain. Professionals from the industry were also asked to participate regarding their large knowledge and experience with advanced technologies. The sample covers

FIGURE 5.1 Research methodology.

TABLE 5.3
Interviewees Profiles

Expert	Experience (Years)	Role
E1	12	Professor and researcher in curriculum design and pedagogy
E2	8	Professor in education
E3	13	Professor in social sciences
E4	10	Educator and researcher in computing
E5	8	Educator and researcher in industrial domain
E6	13	Educator and researcher in industrial domain
E7	20	Education engineering director
E8	7	Engineer and researcher in system engineering
E9	12	Manufacturing manager
E10	25	Consultant in industrial domain
E11	10	Data scientist with focus on education technologies
E12	5	Industrial engineer

perspectives from primary, secondary and higher education from dispersed institutions around the world. Table 5.3 presents the interviewees' experience and roles.

In order to extract knowledge from the interviews, content analysis was performed according to three groups:

1. I5.0 pillars and educational system,
2. Technologies 5.0 and educational system,
3. Challenges of I5.0 and educational system.

5.4 RESULTS AND ANALYSIS

5.4.1 I5.0 Pillars and Educational System

The findings reveal that the integration of I5.0 principles into the educational system will create several opportunities. Interviewees agree that the focus on sustainability, resilience and human centricity will enhance the quality of learning processes and revolutionize the current traditional programmes. Table 5.4 presents the benefits of integration I5.0 pillars with some examples of interviewees' responses.

5.4.2 Technologies 5.0 in Educational System

Interviewees agree that the use of the technologies 5.0 creates several opportunities for the educational system. The contribution of these technologies is summarized in Table 5.5. Expert E10 states: "adopting these tools is no more an option in the current changing environment and the latest pandemic crisis Covid19 has shown how these tools are important to ensure learning continuity".

AI and BDA were the first tools introduced by experts as they allow to analyse and predict students' performance and assist decision-making. In this context expert

TABLE 5.4
I5.0 Pillars and Educational System

		Educational System	Examples
I5.0 pillars	**Sustainability**	Incorporate sustainability focus in educational modules from the elementary school. Establish a research centre for sustainability to align with industry needs Adopt sustainable practices in institutions such renewable energies and waste management systems Held events about sustainability such as workshops and conferences in collaboration with industries.	**E8**: "Incorporating I5.0 principles into educational system will first of all integrate the focus in sustainability into learning programs which will lead to form a conscious generation". **E10**: "Industries can benefit from this integration as they can invest in research center and orient scholars to perform more studies toward sustainable manufacturing". **E1**: "There is a concept that has emerged recently which the ecological schools referring to the one that adopt sustainable practices"
	Resilience	Incorporate resilience focus in educational modules from the elementary school. Develop soft skills such as adaptive and critical thinking. Adopt flexible learning programmes to meet environment changes. Establish recovery strategies and emergency plans to deal with disruptions.	**E2**: "Integrating resilience could help administrators and managers to establish recovery plans and survivability strategies during crisis" **E5**: "The concept of resilience may enhance flexibility of learning programs. During disruptions, the teaching and assessment methods should be adjusted".
	Human centricity	Incorporate human-centric focus in educational modules from the elementary school. Offer customized educational experiences. Foster the user-centred design. Establish partnerships with external organizations to ease the integration of students. Create an accessible and responsive environment for every member of the institution. Promote continuous learning and development for employees. Encourage the innovation and creativity of teachers.	**E7**: "The human centric approach fit very well the educational system as human is already in the center on interest. Therefore, its adoption may enhance the educational experience for educators and students". **E12**: "A greater focus should be devoted to the continuous learning of teachers and students, and also their satisfaction".

TABLE 5.5
Impact of I5.0 Technologies in Educational System

Technology 5.0	Contribution to Educational System	Interviewee
AI/Cognitive AI	AI-powered system analyses student performance AI-based grading systems treat students' submissions Intelligent tutoring system simulates the role of teachers Chatbots timely answer students' questions	All interviewees
Autonomous/ Intelligent Robot	Collaborative robots assist teachers to grade students works Telepresence robots help students to participate in learning process	E1, E6, E7, E8, E9
IoT and sensors	IoT and sensors contribute to building a smart environment for learning Wearable devices monitor students	E1, E5, E6, E9, E10
V/AR	V/AR offer a better learning experience V/AR supports distance learning V/AR ensure interactive and adaptive learning	E6, E9, E11, E12
CS	Secure digital information Responsible for collection and storage of data	E3, E7, E8
CC	Scalable infrastructure Real-time sharing resources Secure and reliable data storage Technologies integration	E7, E11, E9, E2, E5
AM	Facilitate creation of physical prototypes Customized educational tools	E3, E5 E8, E10
CPS	Integrated and connected educational system Optimized and efficient system	E1, E2, E5, E10
HIRT	Track and monitor students' participation in all types of educational activities Promote accessibility and inclusivity for students with specific needs Identification of timely issues	E4, E8, E12

E4 states: "AI algorithms are used to assess students' performance, analyse their behaviours, provide a customized education content and the teaching methods with student needs" and Expert E3 states: "Based on AI, BDA and ML, Intelligent tutoring systems provide virtual lessons and give real-time feedbacks to students. They analyse knowledge gaps and propose suitable programs and activities". He continues: "Another example of these are the AI-based grading systems that grade students' submissions such as essays, quizzes and exams". Expert E12 shed light on the role of chatbots to provide guidance for students: "Chatbots are available 24/24 and can be used anytime to answer students' questions and give guidance in their learning tasks".

IoT, sensors and HITR are used to monitor students and allow more accessibility for students with disabilities. Expert E10 indicates: "Sensors can be integrated to the classroom to enhance the quality of the environment through automatically regulate temperature, lighting, This promotes students and teachers comfort and increase their productivity". Expert E9 adds: "Smart watches are useful for educators and administrators as they provide information about student attendance, engagement and attentiveness to make responsible decisions".

Interviewees emphasis on the contribution of V/AR. Expert E6 states: "V/AR allow to simulate learning environment, helping students to develop their skills safely. As examples, students can explore virtually some geographical area and simulate complex scientific concepts". He adds: "also, it supports distance learning through alternative method different the traditional online courses. For instance, students can use the VR headset to access collaborative discussion and laboratories". Expert E9 declares: "Integrating V/A/MR with ML allows to customize the educational content for student based on their performance and tailored with their lifestyle. One example is the use of AR-powered textbooks that ensure interactivity and adapted learning level".

The successful use of AM in the industry field is also demonstrated in the education domain. Expert E5 states: "3D printers can be used by engineer scholars to easily turn their design ideas into physical models. Equipping Maker spaces with 3D printers and other technologies may encourage students to participate in innovative projects". In the same line expert E8 indicates: "AM can be used to create personalized tools meeting the needs for specific students. For instance, a teacher can 3D print a digestive system model to provide effective explanation".

CC and CS are highlighted to be critical elements for the adoption of technologies 5.0 in the education system. Expert E3 states: "CS is so important for educational system because employees and students' data are sensitive and need to be protected against treats and unauthorized access. In this context, firewalls and encryption protocols are needed". Expert E7 and E11 indicates:

> CC is first saving costs for organizations as it allows to scale resources tailored with students' fluctuations needs and second, it permits to access on-demand to data.
>
> CC is offering a space for students to share their documents, work on their projects and perform collaborative and interactive tasks. It also allows to protect and recover data.
>
> One important thing regarding CC is that is essential for technologies integration.

Finally, the CPS is integrating all the education system 5.0 elements in a connected and optimized world. Expert E5 states: "CPS create integrated and connected educational system that encompasses educators, students, administrators, parents, and also leveraging technologies such as V/AR, IoT and sensors".

5.4.3 Challenges of I5.0 to Education

Although I5.0 has brought several opportunities for education, some challenges are mentioned by experts as follows:

- *Continuous adaptation*: The rapid technological advancement requires a continuous adaptation of the educational system. Expert E6 states: "it is important to mention that keeping up with the rapid evolution of technologies 5.0 is not an easy task". In the same line, expert E12 indicates "The rapid evolution of technologies requires to change methodologies and tools, globally the whole system".

- *Upskilling and Reskilling*: In relation to continuous adaptation, educators need to align their skills with the evolution of technologies. Expert E10 states: "In this new era pedagogical approaches are changing and teachers are asked to follow this evolution to provide better and tailored learning for their students". Expert E6 adds: "Nowadays, students are more conscious and information is available everywhere. Hence, teachers should be able to offer more than traditional learning lessons". Figure 5.2 depicts the educational system in the era of I5.0
- *Professional development*: A focus on technology adoption such as robots may affect the development opportunities for educators. Expert E1 states: "a balance between technology and human centric approaches is pivotal in this context to not ignore the personnel and professional development of educators".
- *Equity concerns*: The access issues of some communities may hinder the large use of technologies 5.0 in an educational system which leads to social inequity. Expert E5 states: "Some educational disparities may arise, if the concerns regarding technology access in some regions are not solved".
- *Lack of collaboration*: The successful implementation of I5.0 into education requires a cross-collaboration between institution and industry. Expert E11 declares: "Developing effective educational programs and tools require

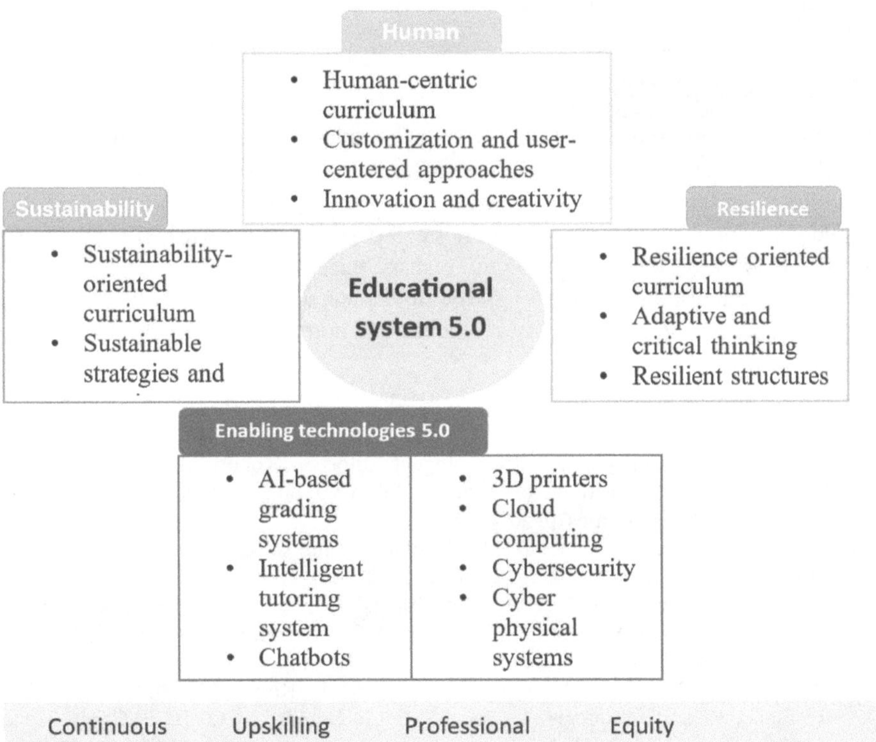

FIGURE 5.2 The educational system in the era of Industry 5.0.

> the involvement of several fields in addition to the social science such as computer science, engineering. Therefore, a collaboration need to be established to fulfil the need for improved learning system".

Mitigating these challenges may require muti-faced strategies as many actors should be involved. A discussion with the participants has resulted in the following key points:

- Make lifelong learning a part of the continuous improvement project of the institutions. To that purpose, flexible learning resources should be accessible and the commitment must be rewarded. Also, integrating the lifelong learning in the curricula as part of the personnel and professional development of the students.
- Develop an agile and responsive curriculum by establishing regular reviews of the learning methodologies, contents, objectives and tools. This requires an investment in the dynamic capabilities to sense and seize the opportunities/threats and reconfigure the ecosystem.
- Foster the academia-industry collaboration through promoting the discussions between experts from both fields. This may lead to offering apprenticeships for students and translating industrial solutions to the education domain.
- From a social perspective, provide support and subsidies for students in areas with low income and propose customized solutions that fit diverse learning needs and conditions.

5.5 CONCLUSIONS

This study explores the integration of I5.0 principles into the educational system. A qualitative approach based on semi-structured interviews was adopted to discuss the ability of I5.0 to improve the learning processes. Twelve experts from education and industry domains involving academics and partitioners were asked to share their knowledge. Findings reveal that I5.0 will reshape the educational system and promote its efficiency and productivity through the implementation of advanced technologies and the integration of sustainability, resilience and human-centricity principles.

REFERENCES

Adel, A., 2024. The Convergence of Intelligent Tutoring, Robotics, and IoT in Smart Education for the Transition from Industry 4.0 to 5.0. Smart Cities 7, 325–369. https://doi.org/10.3390/smartcities7010014

Al-Emran, M., Al-Sharafi, M.A., 2022. Revolutionizing Education with Industry 5.0: Challenges and Future Research Agendas. Int. J. Info. Technol. 6, 1–5.

Alghafri, A.S.R., Ismail, H.N.B., 2014. The Effects of Integrating Creative and Critical Thinking on Schools Students' Thinking. Int. J. Soc. Sci. Humanity 4, 518–525. https://doi.org/10.7763/IJSSH.2014.V4.410

Ali, M.H., Suleiman, N., Khalid, N., Tan, K.H., Tseng, M.-L., Kumar, M., 2021. Supply Chain Resilience Reactive Strategies for Food SMEs in Coping to COVID-19 Crisis. Trends Food Sci. Technol. 109, 94–102. https://doi.org/10.1016/j.tifs.2021.01.021

Alves, J., Lima, T.M., Gaspar, P.D., 2023. Is Industry 5.0 a Human-Centred Approach? A Systematic Review. Processes 11, 193. https://doi.org/10.3390/pr11010193

Bakir, A., Dahlan, M., 2022. Higher Education Leadership and Curricular Design in Industry 5.0 Environment: A Cursory Glance. Dev. Learn. Organ. Int. J. 37, 15–17. https://doi.org/10.1108/DLO-08-2022-0166

Bakkar, M., Kaul, A., 2023. Education 5.0 Serving Future Skills for Industry 5.0 Era. pp. 130–147. https://doi.org/10.4018/978-1-7998-8805-5.ch007

Braun, V., Clarke, V., 2016. (Mis)Conceptualising Themes, Thematic Analysis, and Other Problems with Fugard and Potts' (2015) Sample-Size Tool for Thematic Analysis. Int. J. Soc. Res. Methodol. 19, 739–743. https://doi.org/10.1080/13645579.2016.1195588

Breque, M., De Nul, L., Petridis, A., 2021. Industry 5.0: towards a Sustainable, Human-Centric and Resilient European Industry. Publications Office of the European Union.

Cherrafi, A., Chiarini, A., Belhadi, A., El Baz, J., Chaouni Benabdellah, A., 2022. Digital Technologies and Circular Economy Practices: Vital Enablers to Support Sustainable and Resilient Supply Chain Management in the Post-COVID-19 Era. TQM J. 34, 179–202. https://doi.org/10.1108/TQM-12-2021-0374

Ciolacu, M.I., Alves, G.R., Terkowsky, C., Zoubi, A.Y., Boettcher, K.E.R., Pozzo, M.I., Kist, A.A., 2023. Developing Future Skills in Engineering Education for Industry 5.0: Enabling Technologies in the Age of Digital Transformation and Green Transition, in: Auer, M.E., Langmann, R., Tsiatsos, T. (Eds.), Open Science in Engineering. Springer Nature Switzerland, pp. 1019–1031. https://doi.org/10.1007/978-3-031-42467-0_94

Danaher, J., 2017. Will Life Be Worth Living in a World Without Work? Technological Unemployment and the Meaning of Life. Sci. Eng. Ethics 23, 41–64. https://doi.org/10.1007/s11948-016-9770-5

Danko, C.C., Duarte, A.A.L.S., 2009. The Challenge of Implementing a Student-Centred Learning Approach in Large Engineering Classes. WSEAS Trans. Adv. Eng. Educ.

Ebiere Dorgu, T., 2015. Different Teaching Methods: A Panacea for Effective Curriculum Implementation in the Classroom. Int. J. Second. Educ. 3, 77. https://doi.org/10.11648/j.ijsedu.s.2015030601.13

Echefaj, K., Charkaoui, A., Cherrafi, A., Tiwari, S., Sharma, P., Jabbour, C.J.C., 2024. From Linear to Circular Sustainable Supply Chain Network Optimisation: Towards a Conceptual Framework. Prod. Plan. Control 1–25. https://doi.org/10.1080/09537287.2024.2302479

Ghobakhloo, M., Iranmanesh, M., Tseng, M.-L., Grybauskas, A., Stefanini, A., Amran, A., 2023. Behind the Definition of Industry 5.0: A Systematic Review of Technologies, Principles, Components, and Values. J. Ind. Prod. Eng. 40, 432–447. https://doi.org/10.1080/21681015.2023.2216701

Gürdür Broo, D., Kaynak, O., Sait, S.M., 2022. Rethinking Engineering Education at the Age of Industry 5.0. J. Ind. Inf. Integr. 25, 100311. https://doi.org/10.1016/j.jii.2021.100311

Ivanov, D., 2023. The Industry 5.0 Framework: Viability-Based Integration of the Resilience, Sustainability, and Human-Centricity Perspectives. Int. J. Prod. Res. 61, 1683–1695. https://doi.org/10.1080/00207543.2022.2118892

Kumar, U., Kaswan, M.S., Kumar, R., Chaudhary, R., Garza-Reyes, J.A., Rathi, R., Joshi, R., 2024. A Systematic Review of Industry 5.0 from Main Aspects to the Execution Status. TQM J. 36(6), 1526–1549. https://doi.org/10.1108/TQM-06-2023-0183

Kaswan, M.S., Chaudhary, R., Garza-Reyes, J.A., Singh, A., 2024. A Review of Industry 5.0: from Key Facets to a Conceptual Implementation Framework. Int. J. Qual. Reliab. Manage. (ahead-of-print). https://doi.org/10.1108/IJQRM-01-2024-0030

Leng, J., Sha, W., Wang, B., Zheng, P., Zhuang, C., Liu, Q., Wuest, T., Mourtzis, D., Wang, L., 2022. Industry 5.0: Prospect and Retrospect. J. Manuf. Syst. 65, 279–295. https://doi.org/10.1016/j.jmsy.2022.09.017

Logunova, L.Y., Utkina, A.N., 2019. Human Centricity of Education: Contradictions and Meanings, in: Proceedings of the Middle-Term Conference RC04 Sociology of Education International Sociological Association (ISA), Institute of Foreign Languages

RUDN University, Moscow, 24-26 July 2019. Presented at the Culture and Education: Social Transformations and Multicultural Communication, RUDN University Press, pp. 502–510. https://doi.org/10.22363/09669-2019-502-510

Lozano, R., Merrill, M., Sammalisto, K., Ceulemans, K., Lozano, F., 2017. Connecting Competences and Pedagogical Approaches for Sustainable Development in Higher Education: A Literature Review and Framework Proposal. Sustainability 9, 1889. https://doi.org/10.3390/su9101889

Luz Tortorella, G., Cauchick-Miguel, P.A., Li, W., Staines, J., McFarlane, D., 2022. What Does Operational Excellence Mean in the Fourth Industrial Revolution Era? Int. J. Prod. Res. 60, 2901–2917. https://doi.org/10.1080/00207543.2021.1905903

Maddikunta, P.K.R., Pham, Q.-V., Deepa, P., Dev, N., Gadekallu, K., Ruby, T.R., Liyanage, R., 2022. Industry 5.0: A Survey on Enabling Technologies and Potential Applications. J. Ind. Inf. Integr. 26, 100257. https://doi.org/10.1016/j.jii.2021.100257

McCowan, T., 2016. Three Dimensions of Equity of Access to Higher Education. Comp. J. Comp. Int. Educ. 46, 645–665. https://doi.org/10.1080/03057925.2015.1043237

Mohamed Hashim, M.A., Tlemsani, I., Mason-Jones, R., Matthews, R., Ndrecaj, V., 2024. Higher Education via the Lens of Industry 5.0: Strategy and Perspective. Soc. Sci. Humanit. Open 9, 100828. https://doi.org/10.1016/j.ssaho.2024.100828

Motala, S., 2015. Equity, Access and Quality in Basic Education: A Review. J. Educ. https://doi.org/10.17159/i61a08

Müller, J., 2020. Enabling Technologies for Industry 5.0: Results of a Workshop with Europe's Technology Leaders. Directorate-General for Research and Innovation.

Murphy, C., Carew, P.J., Stapleton, L., 2023. A Human-Centred Systems Manifesto for Smart Digital Immersion in Industry 5.0: A Case Study of Cultural Heritage. AI Soc. https://doi.org/10.1007/s00146-023-01693-2

Nahavandi, S., 2019. Industry 5.0—A Human-Centric Solution. Sustainability 11, 4371. https://doi.org/10.3390/su11164371

Papetti, A., Gregori, F., Pandolfi, M., Peruzzini, M., Germani, M., 2020. A Method to Improve workers' Well-Being Toward Human-Centered Connected Factories. J. Comput. Des. Eng. 7, 630–643. https://doi.org/10.1093/jcde/qwaa047

Rajeev, A., Pati, R.K., Padhi, S.S., Govindan, K., 2017. Evolution of Sustainability in Supply Chain Management: A Literature Review. J. Clean. Prod. 162, 299–314. https://doi.org/10.1016/j.jclepro.2017.05.026

Roy, R., Ateeq, A., 2024. Moving Towards Industry 5.0: Opportunities and Challenges in Bahrain Higher Education, in: The AI Revolution: Driving Business Innovation and Research. pp. 425–435. https://doi.org/10.1007/978-3-031-54379-1_37

Supriya, Y., Bhulakshmi, D., Bhattacharya, S., Gadekallu, T.R., Vyas, P., Kaluri, R., Sumathy, S., Koppu, S., Brown, D.J., Mahmud, M., 2024. Industry 5.0 in Smart Education: Concepts, Applications, Challenges, Opportunities, and Future Directions. IEEE Access 12, 81938–81967. https://doi.org/10.1109/ACCESS.2024.3401473

Villar, A., Paladini, S., Buckley, O., 2023. Towards Supply Chain 5.0: Redesigning Supply Chains as Resilient, Sustainable, and Human-Centric Systems in a Post-Pandemic World. Oper. Res. Forum 4, 60. https://doi.org/10.1007/s43069-023-00234-3

Wan, P.K., Leirmo, T.L., 2023. Human-Centric Zero-Defect Manufacturing: State-of-the-Art Review, Perspectives, and Challenges. Comput. Ind. 144, 103792. https://doi.org/10.1016/j.compind.2022.103792

Xu, X., Lu, Y., Vogel-Heuser, B., Wang, L., 2021. Industry 4.0 and Industry 5.0—Inception, Conception and Perception. J. Manuf. Syst. 61, 530–535. https://doi.org/10.1016/j.jmsy.2021.10.006

Yin, R.K., 2009. Case Study Research: Design and Methods. SAGE.

6 The Role of Industry 5.0 in Improving Sustainability Performance in Manufacturing

Agrifood Case Study

Sana Elhidaoui and Rekha Chaudhary

6.1 INTRODUCTION

I5.0, a cutting-edge industrial framework, combines modern technology with human creativity, encouraging collaboration between human operators and automated systems to address the limitations of I4.0 (Mitra, 2024). The study by Ghobakhloo et al. (2022), which offers a strategic roadmap for achieving sustainability value using I5.0 technologies, illustrates how I5.0 has the potential to enhance sustainability by incorporating human-centric technologies. The objective is to enhance manufacturing efficiency, productivity, and sustainability due to increasing environmental concerns and stringent regulations in the modern manufacturing industry (Elhidaoui et al., 2020). The agrifood sector faces significant environmental and resource pressure due to its significant impact on natural resources. Traditional manufacturing methods often waste energy and materials, necessitating the development of sustainable solutions (Elhidaoui and Kota, 2023).

To create more sustainable and effective operations in the manufacturing environment, I5.0 makes use of cutting-edge technologies like cobots, artificial intelligence (AI), Internet of things (IoT), and advanced data analytics (Rame et al., 2024). While AI and machine learning (ML) help with manufacturing process optimization and predictive maintenance, cobots—which are meant to work alongside human workers—enable increased precision and minimize waste (Sowa et al., 2021). The IoT and smart sensors enable real-time monitoring and control of resources, improving resource management techniques and reducing environmental impact (Soori et al., 2023).

In summary, the shift to I5.0 presents a viable way to address sustainability-related issues as the industrial sector develops further (Kumar et al., 2024). The new Industrial Revolution offers a chance to transform the agrifood industry and promote sustainable practices by promoting mutually beneficial human-machine interaction (Hassoun, 2024).

DOI: 10.1201/9781003535188-6

This chapter investigates the potential of I5.0 to improve sustainability in the agrifood industry by combining human creativity with technological technologies. It includes a case study, implementation methodologies, and best practices to help factories implement I5.0 principles and enhance energy efficiency, waste reduction, and operational efficacy.

6.2 LITERATURE REVIEW

Moving away from automation and data sharing and toward a more human-centric approach, the shift from I4.0 to I5.0 represents a paradigm shift in manufacturing technology (Mitra, 2024). In order to enable the creation of intelligent manufacturing facilities, or "smart factories," the Fourth Industrial Revolution (I4.0) is focused on the integration of the IoT, cyber-physical systems, cloud computing, and cognitive computing (Kagermann and Wahlster, 2022). I5.0 aims to promote cooperation between humans and machines by combining advanced robots with human intelligence. According to Pizoń et al. (2023), this methodology ensures increased production and efficiency while maintaining human values and needs as the top priority. Kumar et al. (2024) emphasize the triple helix model's importance in implementing I5.0, emphasizing the need for collaboration between business, academia, and government. In order to improve human skills without replacing human labor, cobots are made to operate alongside humans (Kaswan et al., 2024). These robots can carry out precise and repeated operations, which reduces the possibility of human error and increases total productivity (Sowa et al., 2021). Because of cobots' collaborative approach, employees may focus on more intricate and creative activities, which eventually increases overall productivity (Javaid et al., 2022).

Optimizing decision-making and manufacturing processes requires the application of AI and ML. AI systems possess the ability to examine vast amounts of data to identify trends and forecast the likelihood of equipment failures (Bangui et al., 2023). By using ML models, production schedules and resource allocation can be optimized and continuously improved, leading to more effective operating procedures (Gardas and Narwane, 2024). IoT devices and smart sensors enable real-time monitoring of various characteristics such as energy consumption, machinery performance, and environmental conditions (Mitra, 2024). Real-time data facilitates swift optimizations and corrective actions, reducing environmental impact and improving resource management (Gardas and Narwane, 2024).

It is often known that conventional production techniques, especially those used in the agrifood industry, frequently lead to significant energy and waste consumption (Elhidaoui et al., 2021). Because of their detrimental effects on the environment and natural resources, these practices are not long-term sustainable (Kanishka and Acherjee, 2023). As environmental restrictions proliferate and customer demand for eco-friendly products rises in tandem, the need for sustainable practices has become more and more critical (Elhidaoui et al., 2024; Sabale et al., 2024).

I5.0 encourages resource efficiency, waste reduction, and energy consumption optimization in order to address the sustainability issues described above (Liu et al., 2024). Pizoń et al. (2023) assert that the amalgamation of advanced technology and human resourcefulness ensures not only maximum productivity but also the ability to modify production procedures in response to changing demands and circumstances. This human-centered approach has particular benefits in the agrifood industry, where sustainability is essential to long-term profitability. The manufacturing industry has effectively implemented I5.0 concepts, as evidenced by a large collection of case studies. The examples above highlight the possibility of significant improvements in sustainability performance, including decreases in energy use, waste generation, and greenhouse gas emissions (Javaid et al., 2022; Liu et al., 2024). Several best practices have been identified by the case studies, such as the use of AI (Heiden et al., 2021) and ML for process optimization (Raven et al., 2024), the integration of cobots for precision tasks (Sivalingam and Subramaniam, 2024), and the deployment of IoT devices for real-time monitoring (Mostofa and Islam, 2023).

6.3 METHODOLOGY

This study uses a qualitative research approach to investigate the impact of I5.0 principles on sustainable performance in the agrifood industry. The case study involves an agrifood company that has adopted I5.0 principles. The analysis provides insights into the company's sustainability procedures, technology use, and results. The study also records challenges faced, best practices, and lessons learned. Semi-structured interviews with professionals in the business, such as managers, engineers, and sustainability employees, are conducted to understand the impact of sustainability, implementation, and technology integration. Thematic analysis is used to identify recurring themes in the interview data set. Academic literature, industry papers, and sustainability evaluations are reviewed for context. Secondary data analysis supports the conclusions drawn from the case study and interviews.

The technique comprises the process of classifying and categorizing qualitative data obtained from interviews in order to uncover key themes related to sustainability and I5.0. Subsequently, these topics are combined to derive more general conclusions on the influence and effectiveness of I5.0 in the agrifood industry. Furthermore, case study outcomes are contrasted to highlight the differences in results due to different technology techniques and business situations, as well as to find common elements that support effective implementation. In addition, pre- and post-I5.0 technology trends in sustainability performance measures are found in order to evaluate the long-term effects of these technologies on sustainability.

Figure 6.1 represents a conceptual diagram illustrating the methodology used in this study.

6.4 FINDINGS

A methodical and strategic approach is required for the deployment of I5.0 in the agrifood industry, with a focus on integrating cutting-edge technologies with human creativity and decision-making processes (Hassoun, 2024). Agrifood firms may significantly improve their sustainability performance by setting clear goals, carrying out a thorough assessment of the current situation, using cobots, applying AI and ML, the IoT, and smart sensors (Mitra, 2024). These tactics can lead to major

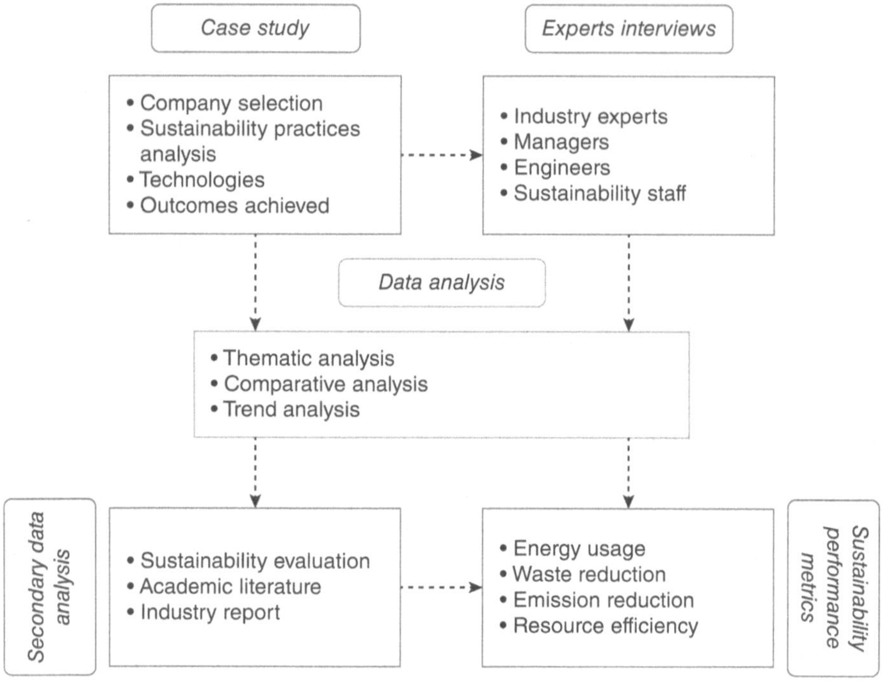

FIGURE 6.1 Methodology framework.

improvements in energy efficiency, waste minimization, and overall operational efficacy through careful planning, continuous observation, and an unrelenting commitment to innovation (Kaswan et al., 2024). This graphic explains how to apply I5.0 principles to increase sustainability in the agrifood industry and provides a clear and simple summary of the main implementation techniques, necessary actions, and anticipated results.

Founded in 1997, Company Alpha is a well-known agrifood firm with a focus on the marketing and processing of seafood. The corporation runs multiple production sites worldwide, with its headquarters located in Morocco. Its market share is especially substantial in Europe and North Africa. The company is dedicated to giving customers wholesome, sustainable, and high-quality food items while reducing its environmental effect. As part of its sustainability efforts, Company Alpha has already installed water-saving devices, recycling initiatives, and energy-efficient lights. However, these steps were insufficient to meet the company's lofty sustainability goals. Consequently, the management team decided to put the I5.0 framework into practice in order to achieve more significant and long-lasting improvements. These are its strategic goals:

- *Improve product quality:* Uphold high standards of product quality while minimizing faults and returns.
- *Reduce waste:* Aim for a 35% waste reduction by improving processes and stepping up all recycling efforts.

- *Minimize energy consumption:* Overall, a 25% reduction in energy consumption within six years.
- *Optimize resource use:* Improve raw material and water efficiency by 20%.

The main goal was to maintain high levels of efficiency and productivity while reducing the company's environmental effect. The main areas of emphasis were maximizing resource usage in agrifood production and minimizing waste and energy consumption.

Notwithstanding its dedication to sustainability, Company Alpha encountered a number of challenges:

- *Resource usage:* The business had trouble making the most use of raw materials and water, which led to an ineffective depletion of these resources.
- *Compliance:* It became essential to undertake quick changes to assure compliance in view of the increasing regulatory pressure for sustainable practices and the concurrent rise in customer demand for environmentally friendly products.
- *Energy consumption:* Due to their high energy usage, the industrial facilities have a large carbon footprint and high operating expenses.
- *Waste generation:* Inadequate processes resulted in considerable waste, comprising organic waste, packaging materials, and by-products.

The organization's specific objectives should be well-defined and supported by quantitative metrics that allow progress to be monitored and measured. One way to do this would be to measure energy use in megawatt-hours (MWh). It is essential to set goals that are feasible given the organization's resources and competencies. Additionally, objectives must be in line with more general organizational strategic goals, like those related to sustainability. Creating a thorough timeline that outlines the necessary steps and dates to complete each goal is crucial. Therefore, it is the responsibility of the relevant parties to assign duties to the appropriate teams and persons in order to ensure accountability and progress tracking.

For the deployment of cobots, it is imperative to identify repetitive and precision-based operations like quality checking, labeling, and packaging. By collaborating with human workers, cobots have the potential to increase productivity and decrease errors in the workplace. The following are the particular duties that were allocated: Cobots have been used for quality control and packaging, among other repetitive, precision-based operations. Waste and human mistakes were decreased as a result of the cobots' integration.

The following was the initiative's result: By using these cobots, production processes were more consistent and efficient, which reduced the number of defective items by 15%. In order to foresee equipment breakdowns and carry out preventative maintenance, AI should be used in predictive maintenance. This strategy might cut down on downtime. By using ML models, production schedules and resource allocation may be optimized, which makes it easier to put more effective operational procedures into place.

By using predictive maintenance, a proactive approach to maintenance can be facilitated by anticipating potential equipment problems. AI algorithms were utilized to predict the likelihood of equipment malfunctions beforehand, which enabled the prompt execution of maintenance procedures and minimized the length of downtime.

Advanced analytical approaches were implemented to optimize the procedure. AI algorithms were employed to optimize production schedules and allocate resources, thereby facilitating more efficient operations. Installation of IoT devices and smart sensors is advised for monitoring energy consumption, environmental conditions, and equipment performance in real time throughout industrial facilities. The amalgamation of sensor data into a centralized system facilitates an all-encompassing examination and prompt remedial measures. With real-time monitoring, variables and processes can be observed quickly, making it easier to spot abnormalities and put remedial measures in place. The goal of installing IoT devices and smart sensors was to track energy consumption, environmental parameters, and mechanical performance in real time. The following was the process's result: These steps made it possible to take remedial action and make improvements right once, which led to a 15% decrease in energy use and a large reduction in greenhouse gas emissions. Data analytics software can be used to track and assess important sustainability parameters, such as water use, trash output, and energy consumption. After determining the waste's origin, specific reduction measures must be put into action. To determine whether the above-described measures are effective, a set of sustainability measurements must be put in place. The application of data analytics technologies made it easier to monitor sustainability measures, identify waste sources, and analyze production data.

The following was the process's result: A 20% decrease in trash output was attained as a result of the data analytics software's actionable insights, which directed the execution of waste reduction programs.

It is advised that staff members receive training on sustainable business practices and the value of their contributions to the accomplishment of company goals. Encouragement of staff participation in the selection and execution of sustainability projects would also be advantageous. Furthermore, it is imperative to demonstrate a leadership commitment to sustainability by means of policy execution, communication tactics, and resource allocation.

It is advised that particular production facility methods be chosen for the I5.0 technology pilot program. A procedure like this would make it possible to determine the possible advantages and difficulties of putting these technologies into practice. The implementation process must be thoroughly documented, along with the difficulties faced and the results obtained. This allows for the sharing of experience and the informed planning of future developments. Furthermore, a study of the pilot projects should be conducted in order to determine the successful tactics and roll them out across the entire organization. At Company Alpha, the application of I5.0 principles led to considerable improvements in sustainability performance. The energy management department submitted the annual energy usage statistics from internal audits carried out between 2020 and 2023, which is one of the data sources. The data set also includes water consumption records from the resource management system with annual efficiency comparisons, waste production logs and annual waste audit reports from the environmental sustainability team, and quality control records detailing defective units and product recalls with monthly and annual summaries from the quality assurance department.

Company Alpha's case study demonstrates the potential of I5.0 in enhancing sustainability performance in the agrifood industry, highlighting improvements in product quality, operational efficiency, waste reduction, energy efficiency, and human-machine collaboration, offering a template for other producers. The agrifood industry can benefit from integrating I5.0 principles and sustainability by integrating human creativity with advanced technologies like cobots, AI, ML, and the IoT. This involves constant adaptation, real-time monitoring for operational efficiency, the use of AI, ML, and data analytics, and human-machine collaboration.

Mostofa and Islam (2023) found that implementing IoT devices and smart sensors resulted in a considerable drop in energy consumption and resource optimization. Furthermore, the integration of AI-driven process optimization and advanced data analytics has reduced waste, hence boosting sustainability (Heiden et al., 2021). However, the process of integrating these technologies is resource-intensive and challenging, requiring a significant initial investment as well as the adaptation of a professional personnel (Ghobakhloo et al., 2022; Liu et al., 2024; Pizoń et al., 2023).

Our results complement and broaden the body of knowledge regarding I5.0 and its effects on sustainability. Furthermore, in line with the findings of Ghobakhloo et al. (2022), which highlighted the value of combining human creativity with cutting-edge technologies like cobots and AI to optimize production processes, our study emphasizes the significance of combining human creativity with technology. However, while earlier studies focused mostly on operational effectiveness, our results provide new perspectives on how these technologies might be applied to cut waste and improve resource use in the agrifood industry.

In contrast to Pizoń et al. (2023), who identified workforce training as a significant obstacle to the adoption of I5.0 technologies, our study revealed that the primary challenge in the agrifood sector was the initial financial investment for the implementation of these technologies, particularly those related to IoT and advanced data analytics. Furthermore, our findings support those of Javaid et al. (2022), who highlighted cobots' good influence on the reduction of production faults. Finally, our findings are similar to the conclusions of Rame et al. (2024). The analysis reveals a significant decrease in energy use, highlighting the potential and problems of I5.0 for sustainability, suggesting future research should bridge conceptual frameworks with actual application.

6.5 CONCLUSION

The I5.0 initiative in the agrifood sector is a significant step toward sustainability, combining human innovation with advanced technologies. This convergence promises increased energy efficiency, reduced waste, and improved product quality. The case study of Company Alpha demonstrates the benefits and challenges, but long-term benefits include regulatory compliance, meeting consumer demand for sustainable products, and gaining a competitive edge. The future of I5.0 in the agrifood sector is promising, with advancements in AI, IoT, and robotics promoting sustainability and widespread adoption. Improved accessibility and collaboration among stakeholders will facilitate the exchange of innovative approaches and best practices.

REFERENCES

Bangui, H., Buhnova, B. and Ge, M. (2023), "Social internet of things: Ethical AI principles in trust management", Procedia Computer Science, Vol. 220, pp. 553–560. https://doi.org/10.1016/j.procs.2023.03.070.

Elhidaoui, S., Benhida, K., Elfezazi, S. and El Hachadi, A. (2020), "Environmental dimension in sustainable supply chain management: Framework and literature review", International Journal of Advanced and Applied Sciences, Vol. 7, No. 8, pp. 74–90. https://doi.org/10.21833/ijaas.2020.08.009

Elhidaoui, S., Benhida, K., Elfezazi, S., Azougagh, Y. and Benabdelhafid, A. (2021), "Model of fish cannery supply chain integrating environmental constraints (AHP and TOPSIS)", Advances in Science, Technology and Engineering Systems Journal, Vol. 6, No. 1, pp. 798–809. https://doi.org/10.25046/aj060188

Elhidaoui, S. and Kota, S. (2023), "Towards a green Agri-food supply chain through ANP and ELECTRE I", Management of Environmental Quality, Vol. 34, No. 5, pp. 1352–1374. https://doi.org/10.1108/MEQ-12-2022-0344

Elhidaoui, S., Kota, S., Benhida, K., and Elfezazi, S. (2024). Industrial green supply chain: A conceptual modelling approach based on artificial intelligence. In Mathiyazhagan, K., Kishore, A., Behdani, B., and Thanki, H. (Eds.), *Impact of industry 4.0 on supply chain sustainability*. Emerald Publishing Limited, pp. 127–142. https://doi.org/10.1108/978-1-83797-777-220241010

Gardas, R. and Narwane, S. (2024), "An analysis of critical factors for adopting machine learning in manufacturing supply chains", Decision Analytics Journal, Vol. 10, p. 100377. https://doi.org/10.1016/j.dajour.2023.100377.

Ghobakhloo, M., Iranmanesh, M., Mubarak, M.F., Mubarik, M., Rejeb, A. and Nilashi, M. (2022), "Identifying industry 5.0 contributions to sustainable development: A strategy roadmap for delivering sustainability values", Sustainable Production and Consumption, Vol. 33, pp. 716–737. https://doi.org/10.1016/j.spc.2022.08.003.

Hassoun, A. (2024). From food industry 4.0 toward food industry 5.0: Human-centric approach for enhanced sustainability and resilience in the food and agriculture sector. In A. Hassoun (Ed.), *Developments in food quality and safety: Food industry 4.0* (Chap. 14). Academic Press.

Heiden, B., Alieksieiev, V., Volk, M. and Tonino-Heiden, B. (2021), "Framing artificial intelligence (AI) additive manufacturing (AM)", Procedia Computer Science, Vol. 186, pp. 387–394. https://doi.org/10.1016/j.procs.2021.04.161.

Javaid, M., Haleem, A., Singh, R.P., Rab, S. and Suman, R. (2022), "Significant applications of cobots in the field of manufacturing", Cognitive Robotics, Vol. 2, pp. 222–233. https://doi.org/10.1016/j.cogr.2022.10.001

Kagermann, H. and Wahlster, W. (2022), "Ten years of industrie 4.0", Science, Vol. 4, No. 3, p. 26. https://doi.org/10.3390/sci4030026.

Kanishka, K. and Acherjee, B. (2023), "Revolutionizing manufacturing: A comprehensive overview of additive manufacturing processes, materials, developments, and challenges", Journal of Manufacturing Processes, Vol. 107, pp. 574–619. https://doi.org/10.1016/j.jmapro.2023.10.024.

Kaswan, M.S., Chaudhary, R., Garza-Reyes, J.A. and Singh, A. (2024), "A review of Industry 5.0: From key facets to a conceptual implementation framework", International Journal of Quality & Reliability Management, Vol. ahead-of-print No. ahead-of-print. https://doi.org/10.1108/IJQRM-01-2024-0030

Kumar, U., Kaswan, M.S., Kumar, R., Chaudhary, R., Garza-Reyes, J.A., Rathi, R. and Joshi, R. (2024), "A systematic review of Industry 5.0 from main aspects to the execution status", The TQM Journal, Vol. 36, No. 6, pp. 1526–1549. https://doi.org/10.1108/TQM-06-2023-0183

Kumar, V., Verma, P., Mittal, A., Gupta, P., Raj, R. and Kaswan, M.S. (2024), "Addressing The kaizen business operations: The role of triple helix actors during COVID-19 outbreak", The TQM Journal, Vol. 36, No. 6, pp. 1665–1694. https://doi.org/10.1108/TQM-08-2023-0253

Liu, S., Li, P., Wang, J. and Liu, P. (2024), "Toward industry 5.0: Challenges and enablers of intelligent manufacturing technology implementation under the perspective of sustainability", Heliyon, Vol. 10 No. 15, p. e35162. https://doi.org/10.1016/j.heliyon.2024.e35162.

Mitra, A. (2024), "Cellular automata-based MapReduce design: Migrating a big data processing model from Industry 4.0 to Industry 5.0", e-Prime - Advances in Electrical Engineering, Electronics and Energy, Vol. 8, p. 100603. https://doi.org/10.1016/j.prime.2024.100603

Mostofa, K.Z. and Islam, M.A. (2023), "Creation of an internet of things (IoT) system for the live and remote monitoring of solar photovoltaic facilities", Energy Reports, Vol. 9, No. Supplement 11, pp. 422–427. https://doi.org/10.1016/j.egyr.2023.09.060.

Pizoń, J., Witczak, M., Gola, A. and Świć, A. (2023), "Challenges of human-centered manufacturing in the aspect of Industry 5.0 assumptions", IFAC PapersOnLine, Volume, Vol. 56 No. 2, pp. 156–161. https://doi.org/10.1016/j.ifacol.2023.10.1562.

Rame, R., Purwanto, P. and Sudarno, S. (2024), "Industry 5.0 and sustainability: An overview of emerging trends and challenges for a green future", Innovation and Green Development, Vol. 3, No. 4, p. 100173. https://doi.org/10.1016/j.igd.2024.100173.

Raven, D.B., Chikkula, Y., Patel, K.M., Al Ghazal, A.H., Salloum, H.S., Bakhurji, A.S. and Patwardhan, R.S. (2024), "Machine learning & conventional approaches to process control & optimization: Industrial applications & perspectives", Computers & Chemical Engineering, Vol. 189, p. 108789. https://doi.org/10.1016/j.compchemeng.2024.108789.

Sabale, D.B., Kaswan, M.S. and Rathi, R. (2024). Development of circular economy model using Industry 4.0's machine learning techniques to improve performance metrics of MSMEs. International Journal of Interactive Design and Manufacturing. https://doi.org/10.1007/s12008-024-01782-6

Sivalingam, C. and Subramaniam, S.K. (2024), "Cobot selection using hybrid AHP-TOPSIS based multi-criteria decision making technique for fuel filter assembly process", Heliyon, Vol. 10, No. 4, p. e26374. https://doi.org/10.1016/j.heliyon.2024.e26374.

Soori, M., Arezoo, B. and Dastres, R. (2023), "Artificial intelligence, machine learning and deep learning in advanced robotics", a review, Cognitive Robotics, Vol. 3, pp. 54–70. https://doi.org/10.1016/j.cogr.2023.04.001.

Sowa, K., Przegalinska, A. and Ciechanowski, L. (2021), "Cobots in knowledge work: Human – AI collaboration in managerial professions", Journal of Business Research, Vol. 125, pp. 135–142. https://doi.org/10.1016/j.jbusres.2020.11.038.

7 Industry 5.0 and Organizational Sustainability

Integrating Cutting-Edge Technologies and Human Ingenuity

Md. Mustaqim Roshid, Abdul Waaje, Rejaul Karim, Mohammad Rafid Irtisham, and Mahender Singh Kaswan

7.1 INTRODUCTION

Industry 5.0 (I5.0) was formally adopted by the European Commission ten years after Industry 4.0 came into being. In the time after, technological change and I5.0 gained significant attention from several sectors of society, including law-makers, scientists, practitioners, and intellectuals (Kishor Kumar et al., 2024). The fifth generation of the industry is here, bringing together cutting-edge digital technology and human creativity to promote better organizational sustainability just as we are on the brink of a distinct paradigmatic change (Dai et al., 2024;; Gobinath et al., 2024; Kumar et al., 2024; Rajkumar et al., 2024). The globe is advancing towards the attainment of sustainable growth targets, with a concentration on establishing durability in the societal, ecological, and economic domains while also promoting loyalty between manufacturers and consumers (Gobinath et al., 2024; Rajkumar et al., 2024). While Industry 4.0 pushes technology forward, it must prioritize preserving ecological balance or improving environmentally friendly practices. Attempts over the last decade to investigate sustainable development using a variety of artificial intelligence (AI) algorithms have fallen short of meeting the challenge of maintaining a consistent and targeted strategy for environmental preservation. Although these technologies—including 3D printing, advanced robotics with AI and ML, the Internet of Things (IoT), data visualization, cloud-based computing, and cyber security—stabilized, the real idea of I5.0 and its rightful place has not yet been established. This is all because of the ongoing Transition from Industry 4.0 to I5.0. According to the CEO and Co-Founder of REInvest Robotics, "Industry 5.0 gives people the things they want and gives them

DOI: 10.1201/9781003535188-7

jobs they are proud of by putting people first in manufacturing." The objective of this revolutionary shift is to achieve ongoing global enhancement by eliminating operations that do not contribute value and reducing waste (Kovari, 2024). The Fifth Industrial Revolution, also known as I5.0, saw the return of human technicians to factory floors to improve the performance of processes (Dai et al., 2024; Gobinath et al., 2024; Kaswan et al., 2024; Rajkumar et al., 2024). However, many academics consented that this was the final stage of Industry 4.0, which had previously relied on cooperation across ever-more-powerful and precise tooling and the distinct imaginative talents of humans.

The shift from Industry 4.0 to I5.0 aims to boost organizational success by sustaining operational excellence in mitigating and accelerating supply chains. This shift also introduces a new level of personalization and a sense of luxury in society that businesses must navigate (Dwivedi et al., 2023; Kumar Varshney et al., 2024; Narkhede et al., 2024; Sahoo et al., 2024). Integrating the Industry 4.0 concept with sustainable manufacturing approaches like Circular Economy and Lean Management is also the demand time in gaining digitalized organizational success towards the path of Sustainable Development Goals (Karim et al., 2024). By integrating sustainability into their operations, companies can improve their long-term profitability, reduce risks, and meet customer, investor, and regulator expectations. Except for SDGs 4, 5, and 16, I5.0 advocates energy savings, they are lowering waste production, and sustainable economic methods for establishing an equilibrium in organizations' impacts on the SDGs, which have wide-ranging applicability (Lozano & Barreiro-Gen, 2023; Narula et al., 2024). From the perspective of contemporary organizational sustainability, this chapter aims to provide a full overview of I5.0 by investigating its definition, evolution, and importance. This presentation will delve into the basic elements and ideas that characterize I5.0, showcasing how the combination of digital technologies and human intelligence can create innovation that is both sustainable and long-lasting.

7.2 OVERVIEW OF INDUSTRY 5.0

I5.0 is currently an increasingly relevant subject as both practical and theoretical innovations persist in progress. The area has consistently been highlighted in social media platforms, institutional study and development programmes, and scholarly works. "Industry 5.0" may sound like a new industrial revolution, provoking questions like "Are we as a society confronting two revolutions simultaneously?" & "If it is an entirely novel revolution, what is the revolutionary technology linked to it?" According to scholars, "Industry 5.0" considers the blending of individual intelligence and cognitive computing technology in working spaces, thus boosting productivity while adding value and prioritizing supervision and managerial duties, greatly restructuring job duties and influencing industrial structures (Kumar Varshney et al., 2024; Narkhede et al., 2024). I5.0 is not directly following Industry 4.0 in time; instead, it expands upon the foundational ideas, innovations, and elements that comprise Industry 4.0 to further eco-friendliness, human-centredness, financial resilience, and personalization through an emphasis on resilience and long-term sustainability. Firstly, the key principles emphasize the importance of human-centricity, which involves placing individuals at the centre of manufacturing operations. The goal is to ensure that technological innovation enhances human skills instead of displacing them. I5.0 requires a socio-technical progression of the human contribution to industrial networks, where every aspect of value chain functions is smartly executed using modern ICT (Panagou et al., 2024; Prassida & Asfari, 2022). Secondly, sustainability for long-lived habitats and social sustainability underpins the fifth industrialization. To achieve social sustainability as a pillar of I5.0, legislation and norms must handle privacy, processing information, human rights, ethics, and psychology (Ciasullo et al., 2024;). The fulfilment of environmental sustainability objectives for the planet and biodiversity, in addition to SDG 11—Smart Cities, highlights the growing concept of a "Society 5.0" that prioritizes human needs and can maintain a balance between revenue generation and ethical behaviour. Thirdly, Resilience, a concept in tandem with SDG 9's commitment to robust infrastructure (Costa, 2024) is a crucial factor in enabling I5.0. Factor in the ability to bounce back from economic and supply chain disturbances, and give sufficient thought to bottlenecks brought on by global epidemics, climatic emergencies, and geopolitical rivalry in manufacturing operations. Fourthly, Personalization, characterized uniquely and beyond customization, has been carrying the weight as a principle of Industry 5. 0. Massive customization enables widespread personalization of items and amenities thanks to substantial technological developments (Abdelhafeez, 2023; Akundi et al., 2022) like digitization, robotization, and digitizing of manufacturing as well as distribution, as well as an extensive knowledge of customer wants and desires. Fifthly, Collaboration across multiple stages of industrial 5.0 notions, such as knowledge management, the human-machine approach, and existing production management concepts like lean manufacturing or circular economy, with the emerging technologies of industry 4.0, is necessary. Industry 4.0 is required for an efficient and robust plant operating process. Collaboration tasks are scheduled or documented in fragmented forms (Narkhede et al., 2024; Narula et al., 2024), and therefore, ranging must be considered to attain optimal utilization potential in the I5.0 framework.

7.3 CUTTING-EDGE TECHNOLOGIES IN INDUSTRY 5.0

7.3.1 Artificial Intelligence and Machine Learning

The development of I5.0 is machine learning and AI, which allows for gradually complex human-machine interactions. With the use of these tools, human-assisted collaborative robots can be developed, increasing productivity and freeing workers to concentrate on creative and analytical tasks. AI-powered results contribute to a more effective and balanced production process by lowering waste, raising sustainability compliance, and optimizing resource use. Machine vision systems, operated by deep learning algorithms, can investigate images or video feeds of products on the production line, identifying defects that might be imperceptible to the human eye (Andrianandrianina Johanesa et al., 2024; Hector & Panjanathan, 2024).

Using great accuracy, highly developed AI algorithms can forecast future demand by analysing massive volumes of data from many sources, such as past sales data, market trends, weather patterns, and social media sentiment (Riccio et al., 2024). This enables producers to maintain product availability while optimizing inventory levels and cutting carrying costs. Machine learning models can also improve logistics operations by recommending the best delivery routes based on traffic patterns, weather, and urgency (Roshid et al., 2024). This lowers transportation expenses while simultaneously enhancing customer satisfaction and delivery times.

7.3.2 Internet of Things (IoT) and Cyber-Physical Systems

The IoT items to the network of physical objects rooted with sensors, software, and other technologies that enable them to collect and exchange data. IoT facilitates the creation of smart factories where machines, products, and even workers are interconnected. According to Mohammed et al. (2024),), and Sheeba et al. (2024), IoT is changing I5.0 by enabling real-time monitoring of production processes, predictive maintenance of machinery, efficient energy management, and enhanced supply chain visibility. With advanced sensors and smart devices, factories can quickly adjust operations, forecast equipment failures, save energy, and track materials seamlessly. This increases productivity, lowers expenses, and supports sustainability initiatives, strengthening industrial operations' flexibility and aligning them with human-centred values. IoT technology is expected to spur more manufacturing innovation as it develops (Mahendrada & Jagdish H Godihal, 2024; Singh et al., 2024; Varghese & Shereef, 2024).

- *Smart Factories:* IoT-permitted devices communicate and collaborate to optimize production processes and energy consumption. Mentions smart factories in I5.0 blend advanced technologies like AI, IoT, and big data with human creativity.
- *Asset Tracking:* Asset tracking in I5.0 uses IoT technologies like smart sensors and RFID tags to observe the location and condition of physical assets in real time.
- *Environmental Monitoring:* This system uses IoT and cloud servers to continuously monitor air quality by tracking harmful gases like NO_2, VOCs, CO, and SO_2 emitted from landfill sites. It also analyses various parameters to understand the causes of pollution, enhancing the overall environmental monitoring and management capabilities in smart cities.

7.3.3 Big Data Analytics and Blockchain

Big Data Analytics and Blockchain are essential technologies in I5.0, driving large transformations in production. Blockchain provides a secure and evident method for recording transactions and data across distributed networks, enhancing supply chain transparency, enabling smart contracts, and ensuring data security (Adekola Adebayo et al., 2024; Akoramurthy et al., 2024; Didwania et al., 2024; Patil & Kulkarni, 2024). Mentions, these technologies facilitate more efficient, transparent, and data-driven manufacturing processes, covering the way for innovation and sustainability in the industrial landscape.

- *Operational Efficiency:* Big data analytics detect inefficiencies in industrial processes by evaluating data from multiple sources. A data-driven strategy makes strategic planning and real-time modifications easier, which lowers costs and increases productivity.
- *Supply Chain Transparency:* Blockchain technology enhances supply chain transparency and traceability by recording transactions on a decentralized ledger. This immutable record reduces fraud and builds trust among stakeholders, allowing for quick identification of issues like counterfeit products.
- *Sustainable Practices:* Data analytics helps companies monitor and reduce environmental impact by tracking resource usage and emissions. Insights from this analysis enable manufacturers to optimize energy consumption and waste management, supporting sustainability goals in I5.0.

7.4 HUMAN FACTORS IN INDUSTRY 5.0

7.4.1 Human-Machine Collaboration

At its core, I5.0 is human-machine Collaboration, the paradigm shift towards smart systems that enable previously unimaginable levels of efficiency and creativity through the interaction of both. Rather than aiming to automate and replace human labour as was the case with other industrial revolutions, the goal of I5.0 is to establish a collaborative setting where machines enhance human abilities (Gobinath et al., 2024; Patil & Kulkarni, 2024). Cobots, or collaborative robots, are created to assist humans in doing complicated jobs and increase output under this paradigm. Thanks to their high-tech sensors and AI algorithms, cobots can comprehend and react to human activities in real time, making the workplace safer and more efficient for everyone (Patil & Kulkarni, 2024).

7.4.2 Enhancing Human Capabilities with Technology

Cutting-edge tech to boost human skills lets employees do things they could not do before. Technology such as virtual reality (VR) and augmented reality (AR) can revolutionize how people experience and engage with both digital and physical environments. Using AR, data and instructions can be superimposed on top of the actual world, giving employees access to information in real time (Brito et al., 2024; Seeliger et al., 2023; Udaya Shankar et al., 2023). Workers may get visual instructions and step-by-step guides right in their line of sight, which is great for training and maintenance situations. When it comes to training and simulation, however, VR creates immersive virtual worlds that enable employees to practice procedures and skills in a risk-free environment. Exoskeletons and other wearable devices augment the human body, allowing workers to lift large objects and do repeated motions with less strain and risk of damage (Alzarrad et al., 2024; Wolfartsberger et al., 2023).

7.4.3 Skills and Competencies for Industry 5.0

Workers must acquire new knowledge and abilities to put I5.0 into action. Workers' skill sets will need to adapt to new technology. Proficiency with advanced manufacturing technology, data analytics, and digital literacy are critical skills for I5.0. Digital literacy covers much ground, from knowing how to use a computer for the most fundamental tasks to being well-versed in cybersecurity and complex programming (Gobinath et al., 2024; Sheela et al., 2024). Employees must be digitally savvy to use digital tools and platforms effectively, comprehend and make sense of data, and keep digital systems secure. Organizations are increasingly relying on data-driven decision-making, making data analytics skills important. Employees need strong skills in data analysis, pattern recognition, and the ability to draw meaningful conclusions (Narula et al., 2024). For this, you will need expert knowledge of data visualization tools, machine learning strategies, and statistical approaches. Creativity, critical thinking, emotional intelligence, and technological prowess are all highly valued in the I5.0 workforce.

7.5 INDUSTRY 5.0 AND ORGANIZATIONAL SUSTAINABILITY

7.5.1 Organizational Sustainability

The capacity of a company to run its daily activities efficiently keeps it afloat in perpetuity while reducing its adverse impacts on people, the planet, and profit is what we mean when we talk about organizational sustainability. By proactively integrating modern technologies and human-centric approaches, organizations can promote creativity, performance, and endurance within the context of I5.0, going beyond straightforward compliance with ecological regulations (Lachvajderová & Kádárová, 2022).

7.5.2 People, Planet, Profit: Triple Bottom Line

Organizational sustainability is based on the triple bottom line (TBL) idea, which states that profit, people, and the environment must all be considered in a balanced manner. Figure 7.1 depicts a TBL.

- *People:* Manufacturing functions in I5.0 prioritize human aspects. Enrich productivity at work, ensure ethical labour, and create a diverse working environment. Corporations can establish less hazardous, more interactive campuses that motivate individuals to focus on innovative and proactive activities using robots and AI-driven solutions (Kemendi et al., 2022).

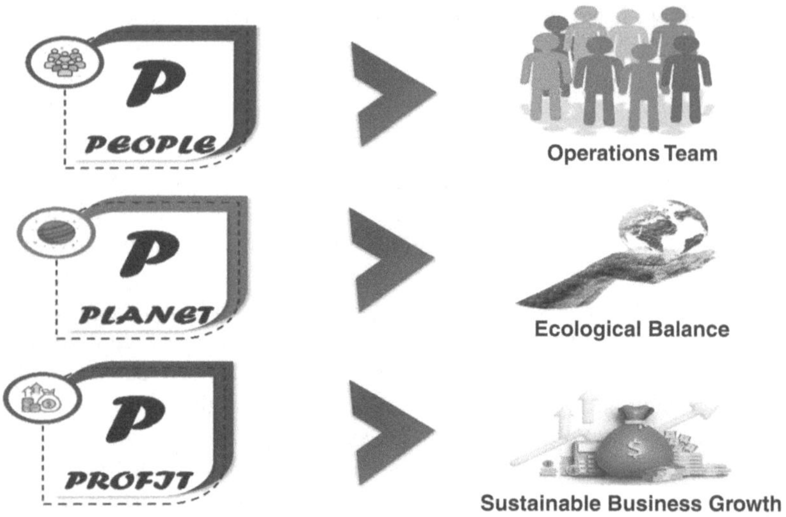

FIGURE 7.1 Triple bottom line.

- *Planet:* TBL emphasizes ecological responsibility. I5.0 promotes green technologies and sustainable practices that mitigate emissions from manufacturing (Karim, 2020). It involves eco-friendly materials, energy from renewable sources, and effective waste handling.
- *Profit:* Financial stability is essential to organizational sustainability which is achieved by ensuring profitability (Karim, Hasan, et al., 2023; Karim, Mamun, et al., 2023;). I5.0 uses sustainable business methods to make money. By adopting digital technologies, enterprises can boost efficiency, cut expenses, and generate new revenue. An integrative strategy achieves fiscal targets while balancing environmental and social obligations (Javaid et al., 2021).

Figure 7.2 depicts a TBL. These eco-friendly procedures rely largely on digital technology that improves various business operations. To begin, the capacity to track and control industrial processes in real time, leading to greater efficiency and less waste, is known as the IoT. In addition, AI can aid sustainability initiatives by analysing data for patterns and providing useful insights (Maddikunta et al., 2022).

7.6 CONCLUSION

A revolutionary step forward, I5.0 emphasizes the complementary nature of cutting-edge digital technology and human imagination. Collaborative robots (cobots) are a key component of the modern workplace, complementing human workers rather than supplanting them. Combined with AR and VR, these cobots allow for

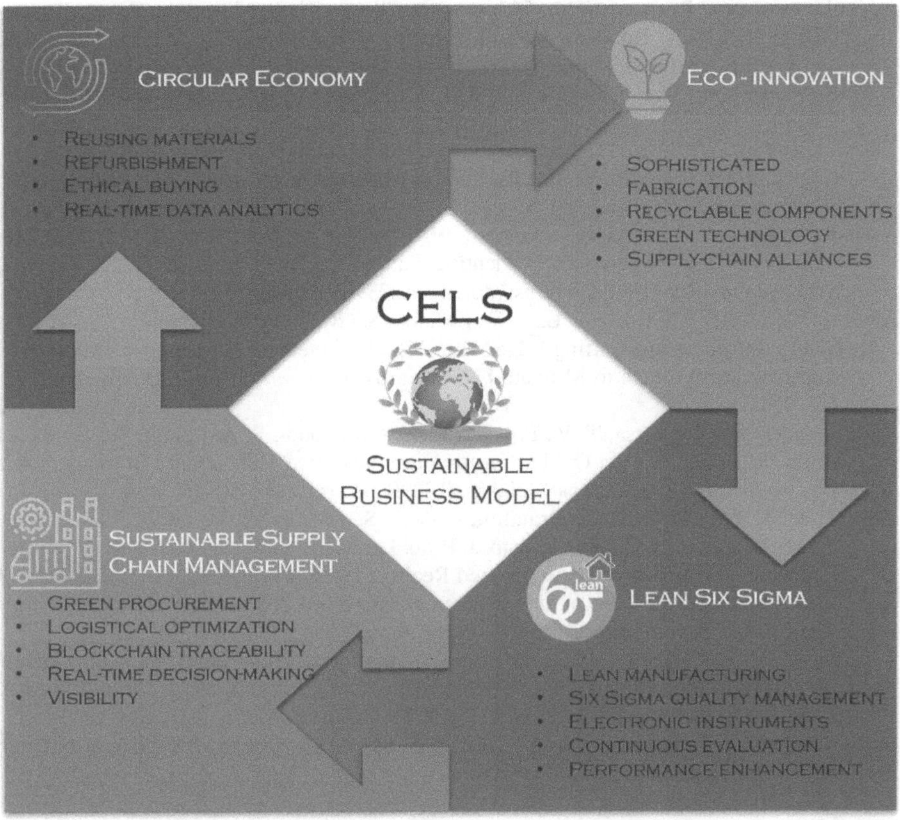

FIGURE 7.2 Sustainable business model.

more immersive training experiences, faster decision-making, and more efficient operations. As a result, innovation and productivity thrive in an environment where people can concentrate on jobs that call for imagination, analysis, and planning. I5.0 has many obstacles, but it also has many potential advantages. By prioritizing sustainability and using circular economy principles, organizations can lessen their negative effects on the environment while improving their operational efficiency. Brand reputation and consumer loyalty can be enhanced by strategically aligning with global sustainability goals, which is not just a statutory necessity but also a business need.

REFERENCES

Abdelhafeez, A. (2023). Industry 5.0: Theoretical Foundations for Enabling Technologies and Applications in Manufacturing Context. *International Journal of Advances in Applied Computational Intelligence*, *4*(1), 37–46. https://doi.org/10.54216/IJAACI.040104

Adekola Adebayo, R., Constance Obiuto, N., Clinton Festus-Ikhuoria, I., & Kayode Olajiga, O. (2024). Robotics in Manufacturing: A Review of Advances in Automation and Workforce Implications. *International Journal of Advanced Multidisciplinary Research and Studies*, *4*(2), 632–638. https://doi.org/10.62225/2583049X.2024.4.2.2549

Akoramurthy, B., Dhivya, K., & Amesho, K. T. T. (2024). Industry Automation: The Contributions of Artificial Intelligence. In *Industry Automation: The Technologies, Platforms and Use Cases* (pp. 57–84).

Akundi, A., Euresti, D., Luna, S., Ankobiah, W., Lopes, A., & Edinbarough, I. (2022). State of Industry 5.0—Analysis and Identification of Current Research Trends. *Applied System Innovation*, *5*(1), 27. https://doi.org/10.3390/ASI5010027

Alzarrad, A., Miller, M., Durham, L., & Chowdhury, S. (2024). Revolutionizing Construction Safety: Introducing a Cutting-Edge Virtual Reality Interactive System for Training US Construction Workers to Mitigate Fall Hazards. *Frontiers in Built Environment*, *10*, 1320175.

Andrianandrianina Johanesa, T. V., Equeter, L., & Mahmoudi, S. A. (2024). Survey on AI Applications for Product Quality Control and Predictive Maintenance in Industry 4.0. *Electronics*, *13*(5), 976. https://doi.org/10.3390/electronics13050976

Brito, E. de M., Arruda, Â. M. de, Brandim, A. de S., Sebaio, A. G., Pereira, B. A., Santos, E. I. da S., Caixeta, H. G., Nascimento, J. P. do, Lemos, L. de F., & Pereira, M. V. (2024). Innovations in Virtual and Augmented Reality: Transforming Organizational Culture Management for the 21st Century. In *Interconnections of Knowledge: Multidisciplinary Approaches*. Seven Editora. https://doi.org/10.56238/sevened2024.010-047

Ciasullo, M. V., La Sala, A., & Ferrara, M. (2024). Normalizing Digitization Into a New 5.0 Equilibrium. Insights from Socio-Ecological Systems. *Springer Proceedings in Complexity*, 107–115. https://doi.org/10.1007/978-3-031-44721-1_9

Costa, E. (2024). Industry 5.0 and SDG 9: A Symbiotic Dance Towards Sustainable Transformation. *Sustainable Earth Reviews*, *7*(1), 1–7. https://doi.org/10.1186/S42055-024-00073-Y

Dai, N. H. P., Giang, P. P., & Kiet, T. H. V. T. (2024). Readiness of Enterprises to Implement Industry 5.0. In *Modern Technologies and Tools Supporting the Development of Industry 5.0* (pp. 408–443). CRC Press.

Didwania, R., Verma, R., & Dhanda, N. (2024). Application of Robotics in Manufacturing Industry. *Machine Vision and Industrial Robotics in Manufacturing*, 57–84. https://doi.org/10.1201/9781003438137-4

Dwivedi, A., Agrawal, D., Jha, A., & Mathiyazhagan, K. (2023). Studying the Interactions among Industry 5.0 and Circular Supply Chain: Towards Attaining Sustainable Development. *Computers & Industrial Engineering*, *176*, 108927. https://doi.org/10.1016/J.CIE.2022.108927

Gobinath, A., Rajeswari, P., Suresh Kumar, N., Anandan, M., & Pavithra Devi, R. (2024). Unleashing the Power of Industry 5.0. In *Using Real-Time Data and AI for Thrust Manufacturing* (pp. 193–211). https://doi.org/10.4018/979-8-3693-2615-2.ch010

Hector, I., & Panjanathan, R. (2024). Predictive Maintenance in Industry 4.0: A Survey of Planning Models and Machine Learning Techniques. *PeerJ Computer Science*, *10*, 1–50. https://doi.org/10.7717/PEERJ-CS.2016/TABLE-2

Javaid, M., Haleem, A., Pratap Singh, R., & Suman, R. (2021). Significance of Quality 4.0 Towards Comprehensive Enhancement in Manufacturing Sector. *Sensors International*, *2*, 100109. https://doi.org/10.1016/j.sintl.2021.100109

Karim, R., Hasan, M., Waaje, A., Yesmin, M. S. H., & Roshid, M. M. (2023). Financial Performance of Bangladeshi Listed Commercial Banks: A Generation-Based Analysis. *International Journal of Finance and Banking Studies*, *12*(4), 46–56. https://doi.org/10.20525/ijfbs.v12i4.3170

Karim, R., Mamun, M. A., Al, & Kamruzzaman, A. S. M. (2023). Cash Conversion Cycle and Financial Performance: Evidence from Manufacturing Firms of Bangladesh. *Asian Journal of Economics and Banking*, *8*(1), 67–82. https://doi.org/10.1108/ajeb-03-2022-0033

Karim, R., Roshid, M. M., Shamme, F. B., & Hasan, M. M. (2023). Non-Performing Loans and Bank Profitability: Evidence from Bangladesh. *International Journal of Finance & Banking Studies*, *11*(4), 95–102.

Karim, R., Roshid, M. M., & Waaje, A. (2024). Circular Economy and Supply Chain Sustainability. In *Strategic Innovations for Dynamic Supply Chains*. https://doi.org/10.4018/979-8-3693-3575-8.ch001

Kaswan, M. S., Chaudhary, R., Garza-Reyes, J. A., & Singh, A. (2024). A Review of Industry 5.0: From Key Facets to a Conceptual Implementation Framework. *International Journal of Quality & Reliability Management* (ahead-of-print). https://doi.org/10.1108/IJQRM-01-2024-0030

Kemendi, Á, Michelberger, P., & Mesjasz-Lech, A. (2022). Industry 4.0 and 5.0 – Organizational and Competency Challenges Of Enterprises. *Polish Journal of Management Studies*, *26*(2), 209–232. https://doi.org/10.17512/pjms.2022.26.2.13

Kishor Kumar, R. C., Pujala, N., & Doss, S. (2024). From Industry 4.0 to 5.0 : Navigating the Human-Tech Symbiosis in the Era of COVID-19. *Ethical Artificial Intelligence in Power Electronics*, 92–104. https://doi.org/10.1201/9781032648323-7

Kovari, A. (2024). Industry 5.0: Generalized Definition, Key Applications, Opportunities and Threats. *Acta Polytechnica Hungarica*, *21*(3), 267–284. https://doi.org/10.12700/APH.21.3.2024.3.17

Kumar Varshney, A., Kumar Singhal, R., Kumar Garg, A., Singhal, R., & Sharma Scholar, H. (2024). The Development of Manufacturing Industry Revolutions from 1.0 to 5.0. *Journal of Informatics Education and Research*, *4*(1), 1526–4726.

Kumar, U., Kaswan, M. S., Kumar, R., Chaudhary, R., Garza-Reyes, J. A., Rathi, R., & Joshi, R. (2024). A Systematic Review of Industry 5.0 from Main Aspects to the Execution Status. *The TQM Journal*, *36*(6), 1526–1549. https://doi.org/10.1108/TQM-06-2023-0183.

Lachvajderová, L., & Kádárová, J. (2022). Industry 4.0 Implementation and Industry 5.0 Readiness in Industrial Enterprises. *Management and Production Engineering Review*, 102–109. https://doi.org/10.24425/mper.2022.142387

Lozano, R., & Barreiro-Gen, M. (2023). Organisations' Contributions to Sustainability. An Analysis of Impacts on the Sustainable Development Goals. *Business Strategy and the Environment*, *32*(6), 3371–3382. https://doi.org/10.1002/BSE.3305

Maddikunta, P. K. R., Pham, Q.-V., Deepa, P., Dev, N., Gadekallu, K., Ruby, T. R., & Liyanage, R. (2022). Industry 5.0: A Survey on Enabling Technologies and Potential Applications. *Journal of Industrial Information Integration*, *26*, 100257. https://doi.org/10.1016/j.jii.2021.100257

Mahendrada, V. S., & Jagdish H Godihal, P. (2024). Transforming Manufacturing: The Power of IoT in Shaping the Future. In *Futuristic Trends in IoT Volume 3 Book 2* (pp. 38–51). Iterative International Publishers, Selfypage Developers Pvt Ltd. https://doi.org/10.58532/V3BIIO2CH5

Mohammed, G. B., Selvarajan, S., Kumar, A. R., Sangeetha, K., Khadidos, A. O., & Vatchala, S. (2024). An Analytical Framework for the Industrial Internet of Things (IIoT): Importance, Recent Challenges, and Enabling Technologies. In *Industry Automation: The Technologies, Platforms and Use Cases* (pp. 1–23). River Publishers. https://doi.org/10.1201/9781003516668-1

Narkhede, G., Chinchanikar, S., Narkhede, R., & Chaudhari, T. (2024). Role of Industry 5.0 for Driving Sustainability in the Manufacturing Sector: an Emerging Research Agenda. *Journal of Strategy and Management* (ahead-of-print).

Narula, S., Tamvada, J. P., Kumar, A., Puppala, H., & Gupta, N. (2024). Putting Digital Technologies at the Forefront of Industry 5.0 for the Implementation of a Circular Economy in Manufacturing Industries. *IEEE Transactions on Engineering Management, 71*, 3363–3374. https://doi.org/10.1109/TEM.2023.3344373

Panagou, S., Neumann, W. P., & Fruggiero, F. (2024). A Scoping Review of Human Robot Interaction Research towards Industry 5.0 Human-Centric Workplaces. *International Journal of Production Research, 62*(3), 974–990. https://doi.org/10.1080/00207543.2023.2172473

Patil, K. P., & Kulkarni, M. S. (2024). Harmonising Human and Robotic Workforces in Industry 5.0. *Modern Technologies and Tools Supporting the Development of Industry, 5*(*0*), 179–208.

Prassida, G. F., & Asfari, U. (2022). A Conceptual Model for the Acceptance of Collaborative Robots in Industry 5.0. *Procedia Computer Science, 197*, 61–67. https://doi.org/10.1016/J.PROCS.2021.12.118

Rajkumar, M., Kapula, P. R., Alekhya, V., Reddy, U., Kumar, A., & Yusupov, A. (2024). Industry 5.0 Revolution. In *Powering Industry 5.0 and Sustainable Development through Innovation* (pp. 43–58). https://doi.org/10.4018/979-8-3693-3550-5.ch004

Riccio, C., Menanno, M., Zennaro, I., & Savino, M. M. (2024). A New Methodological Framework for Optimizing Predictive Maintenance Using Machine Learning Combined with Product Quality Parameters. *Machines, 12*(7), 443. https://doi.org/10.3390/machines12070443

Roshid, M. M., Waaje, A., Meem, T. N., & Sarkar, A. (2024). Logistics 4.0: A Comprehensive Literature Review of Technological Integration, Challenges, and Future Prospects of Implementation of Industry 4.0 Technologies. *International Journal of Technology, Knowledge and Society, 20*(1). https://doi.org/10.18848/1832-3669/CGP/v20i01/65-85

Sahoo, A. K., Alekhya, V., Raj, V. H., Kumar, A., & Shelke, C. (2024). *An In-Depth Analysis of Industry 5.0 in Next-Gen Manufacturing and Differentiations from Industry 4.0* (pp. 154–169). IGI Global. https://services.igi-global.com/resolvedoi/resolve.aspx?doi=10.4018/979-8-3693-3550-5.ch011; https://doi.org/10.4018/979-8-3693-3550-5.ch011

Seeliger, A., Cheng, L., & Netland, T. (2023). Augmented Reality for Industrial Quality Inspection: An Experiment Assessing Task Performance and Human Factors. *Computers in Industry, 151*, 103985. https://doi.org/10.1016/j.compind.2023.103985

Sheeba, A., & Subasini, C. A. (2024, April). Enhancement of Industrial Automation using Internet of Things. In *2024 International Conference on Advances in Data Engineering and Intelligent Computing Systems (ADICS)* (pp. 1–6). IEEE.

Sheela, S. C., Srivastava, S., Kumar, S. V., Aravinda, K., Singh, N., & Yusupov, A. (2024). *Strategic Implementation of Big Data Analytics for Industry 5.0's Sustainable Expansion* (pp. 276–291). IGI Global. https://services.igi-global.com/resolvedoi/resolve.aspx?doi=10.4018/979-8-3693-3550-5.ch019; https://doi.org/10.4018/979-8-3693-3550-5.ch019

Singh, G., Priya, L., Kumar, P., Karthika, J., Saikumar, T., & Nidhya, M. S. (2024). Analysis of Internet of Things-Integrated Technology in a Smart Factory. In *Machine Vision and Industrial Robotics in Manufacturing* (pp. 368–384). CRC Press. https://doi.org/10.1201/9781003438137-20

Udaya Shankar, A., Tewari, V., Rahman, M., Mishra, A., & Bajaj, K. (2023). Impact of Virtual Reality (VR) and Augmented Reality (AR) in Education. *Tuijin Jishu/Journal of Propulsion Technology, 44*(4), 1310–1318. https://doi.org/10.52783/TJJPT.V44.I4.1014

Varghese, N., & Shereef, S. (2024). Scope of Industrial Internet of Things in Manufacturing Industry: Challenges, Recent Trends and Applications. In *Futuristic Trends in Artificial Intelligence Volume 3 Book 2* (pp. 161–173). Iterative International Publishers, Selfypage Developers Pvt Ltd. https://doi.org/10.58532/V3BDAI2P2CH9

Wolfartsberger, J., Zimmermann, R., Obermeier, G., & Niedermayr, D. (2023). Analyzing the Potential of Virtual Reality-Supported Training for Industrial Assembly Tasks. *Computers in Industry*, *147*, 103838. https://doi.org/10.1016/J.COMPIND.2022.103838

8 Lean Service 5.0: A Guide to the Production Management Strategies

A Proposal Based on Systems Theory

David de Oliveira Costa, Andrei Bonamigo, and Davide Calandra

8.1 INTRODUCTION

In the context of complex organizational environments, strategic alignment is crucial for effective decision-making in production management. The influence of organizational culture in this process can influence the perception and interpretation of the strategies adopted at the level of processes and operations [1]. In this sense, leadership plays a fundamental role in guiding decisions based on concrete data, ensuring that strategic choices are aligned with organizational reality, and minimizing possible conflicts arising from cultural differences [2, 3]. As identified in their research [4], the authors identified a statistically significant relationship between the decision-making style and the dimensions of national culture in the production system.

The author in [5] reports the challenge of establishing a structure focused on serving and fulfilling the business vision due to the need to link strategic planning to operational management. Understanding that deploying specific actions is crucial for maintaining organizations based on lean and continuous improvement in production management. Another aspect points to the need to align the behaviors and attitudes of the leadership team as a fundamental element. The authors in [6] point to the elements of tools and practices that Lean Service encompasses and that the human element is crucial in this contour, with improvements in customer satisfaction and operational costs seen through such practices, but forget it is important to emphasize that, at an earlier stage, if there is no strategic structuring and planning based on a reliable database, these practices may not have a significant effect on the business and strategies employed in production management.

The objective of this chapter is to develop a structured framework that integrates corporate strategy with Lean Service 5.0. This integration aims to maintain strategic alignment and enhance agility in decision-making with precision and confidence. The proposed framework includes the incorporation of human development aspects and interconnected technological foundations (Industry 5.0), emphasizing

DOI: 10.1201/9781003535188-8

customer perception and perceived [7]. This approach addresses the current imperative to generate competitive advantage in the market through production management strategies [8].

This chapter boils down to understanding three categories portrayed with research problems: How do the principles of LS5.0 influence the organizational culture of a business and how does it affect the dynamics of decision-making in a corporation? Having a clear and defined understanding that relevant themes in the sphere of strategic decision-making will depend on the company's senior leadership; therefore, organizational culture factors should reflect on behavior, as well as aspects of the corporate vision should form part of your reasoning map.

Based on this understanding, the respective research questions were structured, as follows: How does the LS5.0 concept influence the company's organizational culture and how does this affect the decision-making dynamics in a corporation? How can the LS5.0 framework shape the mindset of organizational culture and influence top management from data-driven decision-making in the production management perspective? Finally, what are the production management challenges that companies face when adopting LS5.0 principles?

8.2 METHODOLOGY

Our chapter aims to present a proposal for integration between elements of Lean Service 5.0, Lean Thinking, and decision-making process requirements to strengthen the corporate strategy together, as shown in Figure 8.1. The initial studies resulted in the structuring of the research, and, as a result, we obtained the methodological flow of this chapter (Figure 8.1) and a proposal for an integrated framework for the respective object of study. Given the proposal, the search string was used: "corporate strategy" and lean OR "lean manufacturing" OR "lean service" OR "lean thinking"

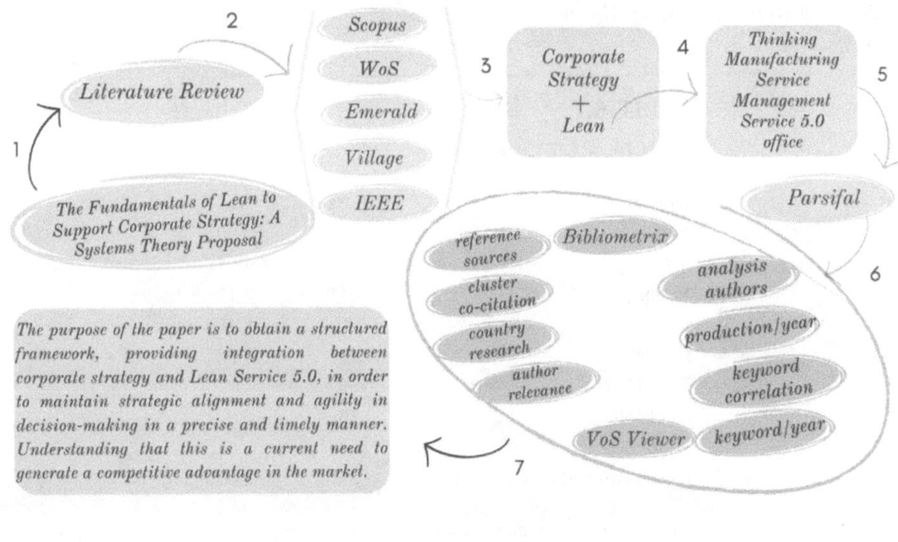

FIGURE 8.1 Paper methodological flow. (Source: Authors' elaboration.)

OR "lean management" OR "lean service 5.0" OR "lean office". This string follows the respective structure: the first stage is the population, followed by the intervention, and finally, the objective, based on the PICOC methodology [9].

This structuring and arrangement were based on proposal and content analysis. Thus, the pre-analysis was considered the stages of planning and methodological approach, followed by the literature research stage with documents from databases validated in academia and, finally, the analytical phase to direct the respective research [10]. Step 1: Identify the research question and define the objective of the study; Step 2: Identify relevant studies; Step 3: Select pertinent studies; Step 4: Map the data; and, Step 5: Report the results.

After defining the central proposal of this research (7), and associating it with the derivations of lean and corporate strategy (3; 4), it was understood the need to, together with the search string in bases with academic relevance (2), structure an adequate literature review (1). Thus, with the support of the Parsifal platform [11] (5), a stratification of papers was obtained to add to the research. With this refinement, the VOSviewer [12] and Bibliometrix platforms [13] were used to obtain a robust bibliometric base (6).

8.3 RESULTS

Applying the Parsifal computational platform [28], a resource to support the structuring of a systematic literature review, seeking to consider only documents with academic representation and scientific recognition in this research, the Proknow-C methodology was applied [31], where the cutoff point was the first papers with high citation (Table 8.1).

The intense interaction of this topic among many countries demonstrates the relevance, impact, and interest in this approach (Figure 8.2).

The journals with the greatest coverage were found based on the respective bases, according to the proposed theme (Figure 8.3).

TABLE 8.1
Papers × Database

Database	Papers	%
WoS	780	65.99
Scopus	58	4.91
Village	25	2.12
IEEE	100	8.46
Emerald	219	18.53
Total Global	1182	
Duplicates	18	
Valid	1164	
Papers Aligned to Research	31	2.66

Source: Authors' elaboration.

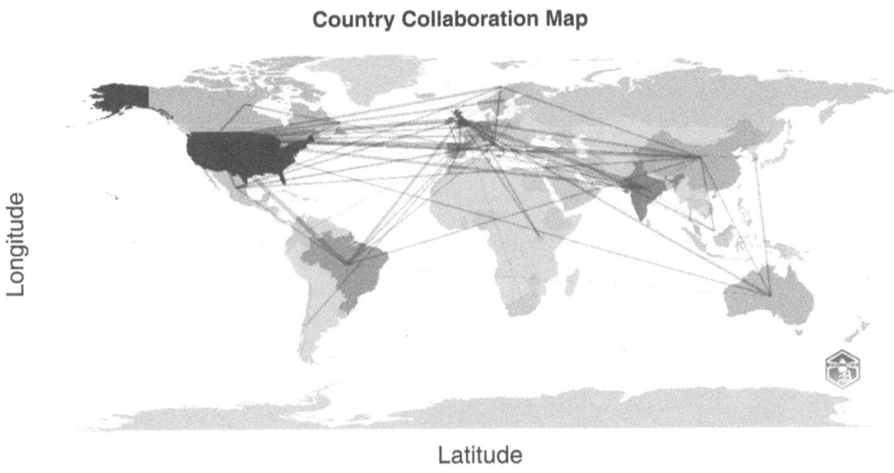

FIGURE 8.2 Analysis of interaction between countries of the research topic. (Source: Authors' elaboration.)

Thus, according to the respective database, there are 29 authors who met this requirement. It is clear, when analyzing the respective bases, the formation of clusters of research groups (Figure 8.4), that there is strong collaboration between researchers in their groups, but there is no evidence of interrelation between the different groups.

Given the papers found in the respective databases, they were analyzed in detail regarding their adherence to the aforementioned research questions to score points based on alignment. Thus, if the paper fully aligns with the respective questions, the score will be 1. If it partially answers, the equivalent score will be 0.3. Finally, if there is no equivalence to the questions, the score will be 0. Ultimately, these points will be added together, and each paper will have an overall score. The higher this score, the greater adherence to the research object (Table 8.2).

After analyzing the document portfolio, it can be stated that only the index of adherence between the findings in the literature and the research object is less than 1%. To highlight the rate of use of findings in the literature at each stage of the research, the structure proposed [14] (Figure 8.5).

Based on the methodological proposal that this research followed, by prioritizing papers with greater academic prestige (citation index), Table 8.3 is expressed in a descending ranking. Notably, the paper "Lean Management as Part of the Corporate Strategy" was not analyzed due to the unavailability of access via the CAPES repository and the ResearchGate platform.

8.4 DISCUSSION

The inputs obtained from the constructs of this literature review allowed structuring, based on its main elements, the interrelation between aspects of Lean Thinking and corporate strategy, obtaining a rational database for adequate decision making, but

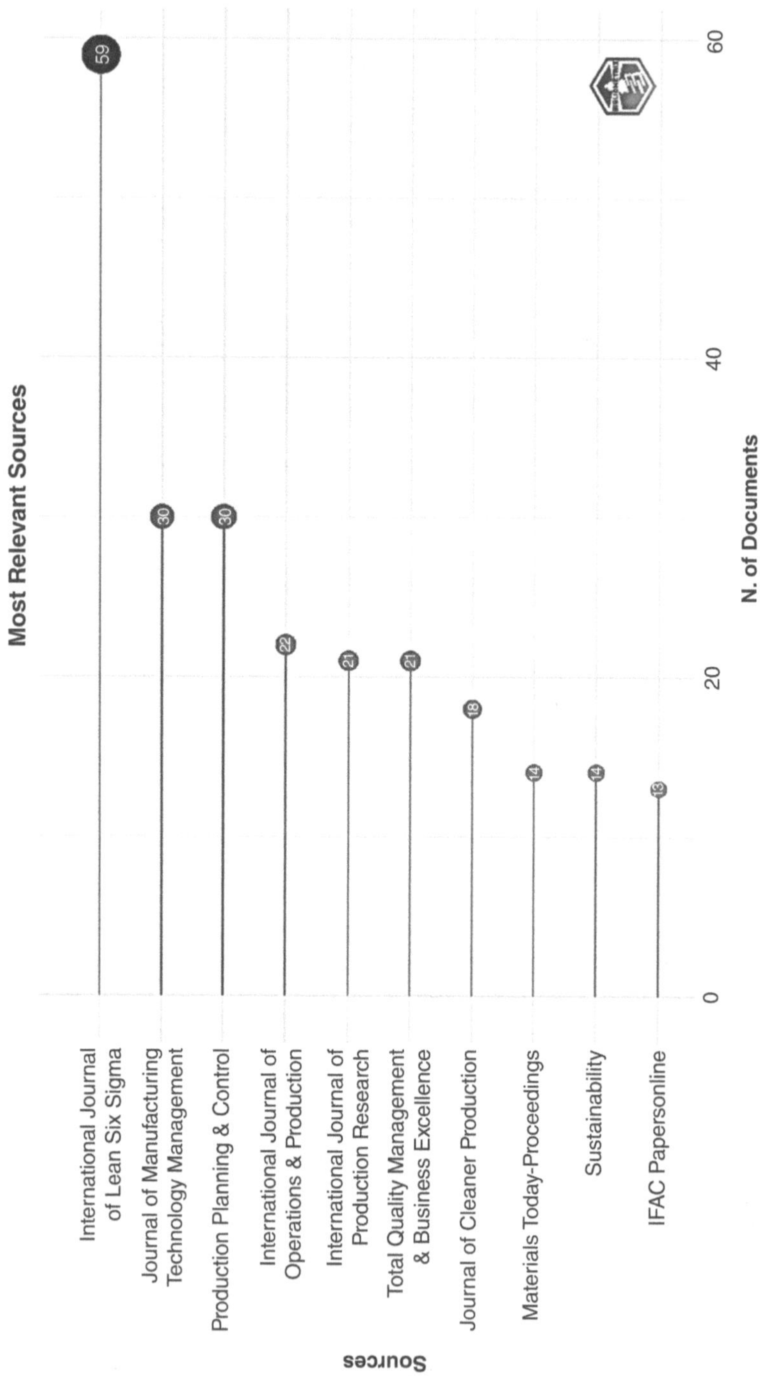

FIGURE 8.3 List of journals relevant to the topic. (Source: Authors' elaboration.)

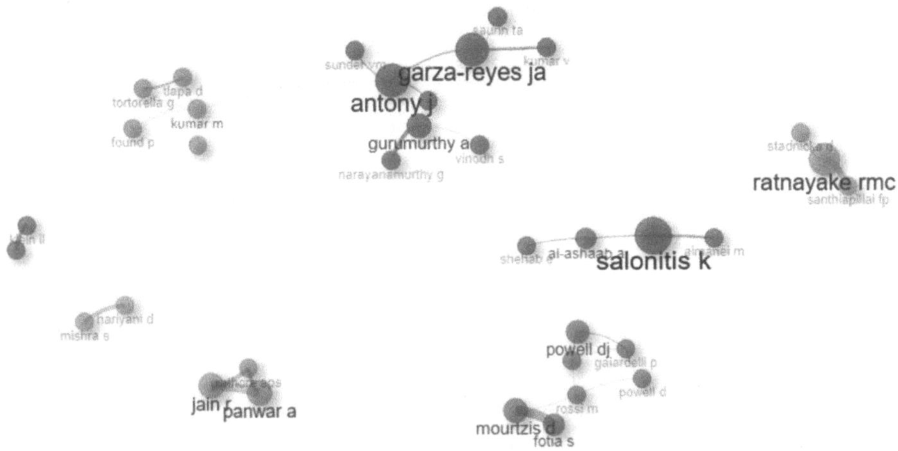

FIGURE 8.4 Researcher group clusters. (Source: Authors' elaboration.)

contemplating the organizational structure (respect for people as a structural axis), aligned with the LS5.0 proposal. Thus, this chapter enabled a deeper understanding of the integration between LS5.0 principles and corporate strategy, thus promoting organizational development and reinforcing the proposition of this framework (Figure 8.6).

TABLE 8.2
Papers Aligned to Research Questions

Author	Influence of LS5.0 on Organizational Culture and Decision-Making Dynamics	Challenges Companies Face When Embracing LS5.0 Principles	Impact of LS5.0 on Organizational Culture Mindset and Data-Driven Decision-Making	Sum
Coetzee, R. et al. (2016)	1	1	1	3
Tortorella, G. et al. (2021)	1	0.3	1	2.3
Loyd, N. et al. (2020)	1	1	0.3	2.3
Helmold, Marc (2020)	1	1	0.3	2.3
Netland, T. H. et al. (2015)	1	1	0	2
Freitas, R. et al. (2015)	1	0.3	0.3	1.6
So, Stuart and Sun, Hongyi (2015)	0.3	0.3	1	1.6
Romero, D. et al. (2019)	0	1	0.3	1.3
Al-Ghamdi, S. M. et al. (2007)	0.3	1	0	1.3
Ghobakhloo, M. and Fathi, M. (2020)	0	0	1	1

Source: Authors' elaboration.

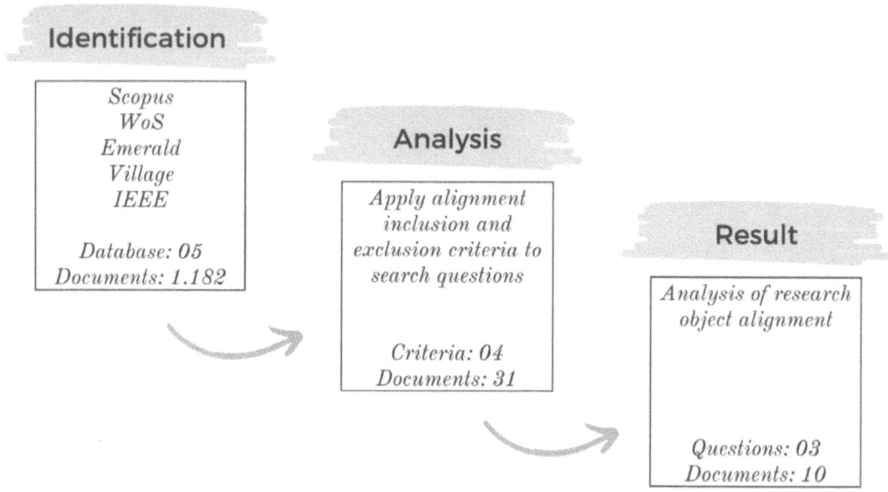

FIGURE 8.5 Research fundamentals. (Source: Authors' elaboration.)

TABLE 8.3
Research Construct

Paper	Structural Pillars
[15]	Employee Engagement – Commitment Levels – Cultural Change – Employee Buy-In
[16]	Alignment of Values – Supportive Leadership – Employee Engagement – Training and Development – Continuous Communication – Performance Measurement
[17]	Commitment and Engagement – Motivation and Buy-In – Communication and Collaboration – Respect for People – Continuous Improvement Mindset
[18]	Employee Engagement – Teamwork and Collaboration – Knowledge and Skills – Ownership and Accountability – Continuous Improvement Mindset
[19]	Efficiency and Productivity – Knowledge Creation and Transfer – Iterative Learning – Problem-Solving Skills – Feedback Mechanisms
[20]	Process Focus and Streamlining – Delegation and Knowledge Development – Continuous Improvement with Kaizen Methodology – Leaner Organizational Model – Empowerment and Training of Workforce
[21]	Continuous Learning and Improvement – Process-Centric Approach – Efficiency Gains and Value Creation – Process-Culture-Technology Alignment – Customer-Centricity and Profitability
[22]	Effective Communication Strategies – Employee Empowerment – Continuous Improvement Mindset – Leadership Commitment
[23]	High-Performance Mindset – Flexibility and Adaptability – Quality and Innovation – Cost Reduction and Waste Elimination – Competitive Advantage

Source: Authors' elaboration.

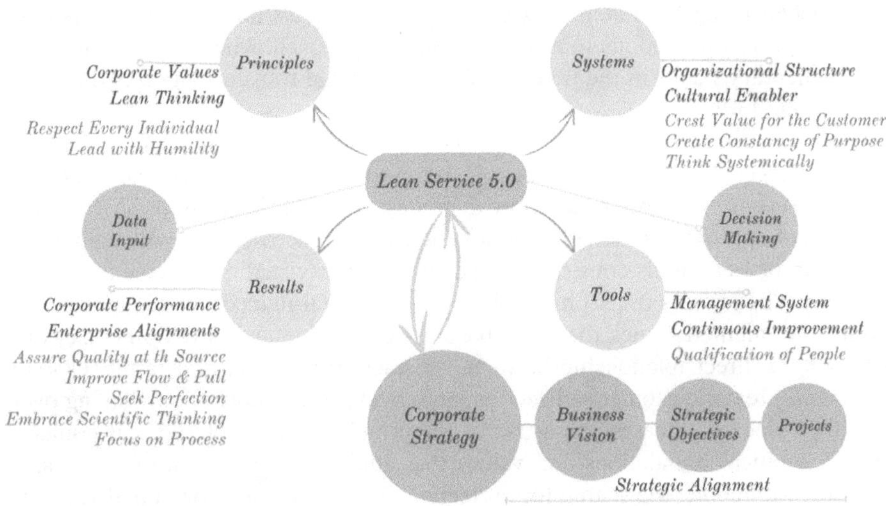

FIGURE 8.6 Integrative framework. (Source: Authors' elaboration.)

The authors in [24] consider the Lean philosophy to highlight, with practical situations, the versatility and scope of this thought. In this understanding, we seek to apply the proposal established in this framework, starting from the corporate strategy and the need to establish a strategic alignment so that the established projects are aligned with the business needs and that this contributes to the established vision. Therefore, the database, seeking to understand the bottlenecks and critical points of the business, must be reliable. Consequently, this corporate organization must have people prepared, qualified, and able to perform their respective functions [12, 25], and, given the composition of the scenario, people need to know how to solve complex problems; this skill is vital for business sustainability and a culture of excellence. Starting with leadership commitment and involvement, this is a key component in establishing a culture of continuous improvement and innovation, showing a significant impact on organizational performance and employee engagement [26]. There are many methods and strategies to produce the best decision. The authors in [27] stated that respecting people does not just mean treating them with courtesy but also involves recognizing their capabilities, valuing their contributions, and giving them autonomy to make decisions that directly affect their work.

By integrating structural elements and practices of Lean Thinking, the proposal of this framework aims to establish a connection with the corporate strategy to align actions with an emphasis on the business vision of what will add perceived value [27, 28]. In this understanding, the authors in [29] unfolded these principles into practical objectives: Value Stream Mapping (VSM), to identify opportunities for improvements and identification of the bottleneck, Heijunka and Kanban as an element of leveling and balancing in production lines, 5S a philosophy of respect for people, when using each step, it is clear that the benefits aim to preserve people. The Lean philosophy points to a socio-technical approach, where the human being is an elementary factor in the process of organizational transformation and process improvement [30].

The authors in [31] summarized this structure into a rational division; in the tools approach stage, the focus is on specific techniques used in a practical approach, whether to eliminate waste or to improve processes (VSM, Kanban, or SMED). The next level of approach is the creation of systems and processes that will support the application of Lean tools consistently and sustainably (visual management systems, TPM, 5S). The authors in [32] highlighted that it is vital in a corporate environment to introduce methodologies that seek business excellence and develop people, thus identifying a direct relationship between satisfactory performance, faster response to changes, greater customer satisfaction, and greater team motivation. Having overcome the respective phases, we arrive at the principles approach; here, it points to the philosophical foundations and values that guide the application of Lean, such as respect for people, the search for perfection, and continuous flow. Finally, results refer to the benefits achieved by the effective implementation of Lean, such as cost reduction, quality improvement, increased productivity, and greater customer and employee satisfaction.

The authors in [33] understand that these pillars constitute the foundation for the successful implementation of Lean in organizations, always seeking to eliminate waste and create value continuously. This will be related to the LS5.0 proposal, where not only the systemic approach to tools or philosophies but also the interconnection of these themes with the corporate strategy have necessary inputs to position rational and coherent decision-making to the established scenario [12]. This strategic perspective addresses long- and short-term strategic actions, in order to guide the direction of competitive strategies based on production management [28, 34].

8.5 CONCLUSION

This chapter aims to develop a structured framework that integrates corporate strategy with LS5.0 to insert social-technical aspects in this context. Understand that this framework is a humanized rational structure. Based on the results found in this study, within the pre-established objectives, when thinking about the relationship between LS5.0 and corporate strategy, each research question was properly thought out as it would be approached in practice at all corporate levels. Thus, it was possible to develop this framework, which allows to interconnect, in a systemic approach, elements of Lean Thinking through the structural principles of Lean, the stage focused on systems where the organizational structure and corporate culture are immersed, the essence of Lean tools that support and effectively constitute the structural apex for business results. This situation will support the corporate strategy with defined objectives and projects identified as potential bottlenecks critical to the pre-established vision.

Based on this framework, it was possible to understand corporate dynamics to prioritize strategic actions and projects vital to the business, rationalizing available resources with effectiveness and focus.

REFERENCES

1. P. E. Johansson, J. Bruch, K. Chirumalla, C. Osterman, and L. Stålberg, "Integrating advanced digital technologies in existing lean-based production systems: Analysis of paradoxes, imbalances and management strategies," *International Journal of Operations & Production Management*, vol. 44, no. 6, pp. 1158–1191, May 2024. doi: 10.1108/IJOPM-05-2023-0434.
2. V. van Loenen, R. Schouteten, M. Visser, and E. Vosselman, "The translation of Lean management: Prospects of a relational approach for successful practice," *Public Money & Management*, vol. 44, no. 1, pp. 54–61, Jan. 2024. doi: 10.1080/09540962.2022.2111884.
3. A. Bonamigo, P. Arcanjo, M. Goncalves, N. Pereira, and D. Cunha da Silveira, "Lean 4.0 in port management: An alternative to support the development of the circular economy in the sector," *Acta Logistica*, vol. 10, no. 2, pp. 291–304, Jun. 2023. doi: 10.22306/al.v10i2.395.
4. M. Dabić, D. Tipurić, and N. Podrug, "Cultural differences affecting decision-making style: A comparative study between four countries," *Journal of Business Economics and Management*, vol. 16, no. 2, pp. 275–289, Dec. 2014. doi: 10.3846/16111699.2013.859172.
5. M. Helmold, "Lean Management as Part of the Corporate Strategy," 2020, pp. 45–55. doi: 10.1007/978-3-030-46981-8_5.
6. F. Ackermann, C. Eden, and P. McKiernan, "Stakeholders in strategy-making," *Journal of Strategy and Management*, vol. 17, no. 2, pp. 282–296, May 2024. doi: 10.1108/JSMA-11-2023-0280.
7. S. K. C. De Azevedo, L. C. Ferreira, H. M. Rocha, and A. Bonamigo, "Service 5.0: Conceptual basis and proposal of standardized terminology," *CONTRIBUCIONES A LAS CIENCIAS SOCIALES*, vol. 17, no. 2, p. e5361, Feb. 2024. doi: 10.55905/revconv.17n.2-280.
8. T. V. Pogodina, T. V. Muzhzhavleva, and N. L. Udaltsova, "Strategic management of the competitiveness of industrial companies in an unstable economy," *Entrepreneurship and Sustainability Issues*, vol. 7, no. 3, pp. 1555–1564, Mar. 2020. doi: 10.9770/jesi.2020.7.3(9).
9. T. F. Galvão, and M. G. Pereira, "Revisões sistemáticas da literatura: passos para sua elaboração," *Epidemiologia e Serviços de Saúde*, vol. 23, no. 1, pp. 183–184, Mar. 2014. doi: 10.5123/S1679-49742014000100018.
10. L. Bardin, *Análise de conteúdo*. Edições 70, 2011. Accessed: Jun. 17, 2024. [Online]. Available: https://madmunifacs.wordpress.com/wp-content/uploads/2016/08/anc3a1lise-de-contec3bado-laurence-bardin.pdf
11. B. Kitchenham, and P. Brereton, "A systematic review of systematic review process research in software engineering," *Information and Software Technology*, vol. 55, no. 12, pp. 2049–2075, Dec. 2013. doi: 10.1016/j.infsof.2013.07.010.
12. N. J. van Eck, and L. Waltman, "Software survey: VOSviewer, a computer program for bibliometric mapping," *Scientometrics*, vol. 84, no. 2, pp. 523–538, Aug. 2010. doi: 10.1007/s11192-009-0146-3.
13. M. Aria, and C. Cuccurullo, "bibliometrix : An R-tool for comprehensive science mapping analysis," *Journal of Informetrics*, vol. 11, no. 4, pp. 959–975, Nov. 2017. doi: 10.1016/j.joi.2017.08.007.
14. A. Toth-Peter, R. Torres de Oliveira, S. Mathews, L. Barner, and S. Figueira, "Industry 4.0 as an enabler in transitioning to circular business models: A systematic literature review," *Journal of Cleaner Production*, vol. 393, p. 136284, Mar. 2023. doi: 10.1016/j.jclepro.2023.136284.
15. T. L. Doolen, and M. E. Hacker, "A review of lean assessment in organizations: An exploratory study of lean practices by electronics manufacturers," *Journal of Manufacturing Systems*, vol. 24, no. 1, pp. 55–67, Jan. 2005. doi: 10.1016/S0278-6125(05)80007-X.

16. G. L. Tortorella, D. Fettermann, F. S. Fogliatto, M. Kumar, and D. Jurburg, "Analysing the influence of organisational culture and leadership styles on the implementation of lean manufacturing," *Production Planning & Control*, vol. 32, no. 15, pp. 1282–1294, Nov. 2021. doi: 10.1080/09537287.2020.1799255.
17. N. Loyd, G. Harris, S. Gholston, and D. Berkowitz, "Development of a lean assessment tool and measuring the effect of culture from employee perception," *Journal of Manufacturing Technology Management*, vol. 31, no. 7, pp. 1439–1456, Mar. 2020. doi: 10.1108/JMTM-10-2019-0375.
18. T. H. Netland, J. D. Schloetzer, and K. Ferdows, "Implementing corporate lean programs: The effect of management control practices," *Journal of Operations Management*, vol. 36, no. 1, pp. 90–102, May 2015. doi: 10.1016/j.jom.2015.03.005.
19. R. C. Freitas, M. C. D. Freitas, G. Gomes de Menezes, and R. S. Odorczyk, "Lean office contributions for organizational learning," *Journal of Organizational Change Management*, vol. 31, no. 5, pp. 1027–1039, Sep. 2018. doi: 10.1108/JOCM-06-2017-0221.
20. S. So, and H. Sun, "Lean thinking as organisational practice in enabling supply chain sustainability," *International Journal of Environmental Technology and Management*, vol. 18, no. 4, p. 291, 2015. doi: 10.1504/IJETM.2015.072140.
21. D. Romero, M. Flores, M. Herrera, and H. Resendez, "Five Management Pillars for Digital Transformation Integrating the Lean Thinking Philosophy," in *2019 IEEE International Conference on Engineering, Technology and Innovation (ICE/ITMC)*, IEEE, Jun. 2019, pp. 1–8. doi: 10.1109/ICE.2019.8792650.
22. S. M. Al-Ghamdi, M. H. Roy, and Z. U. Ahmed, "How employees learn about corporate strategy," *Cross Cultural Management: An International Journal*, vol. 14, no. 4, pp. 273–285, Oct. 2007. doi: 10.1108/13527600710830313.
23. M. Ghobakhloo, and M. Fathi, "Corporate survival in industry 4.0 era: The enabling role of lean-digitized manufacturing," *Journal of Manufacturing Technology Management*, vol. 31, no. 1, pp. 1–30, Jul. 2019. doi: 10.1108/JMTM-11-2018-0417.
24. A. Chiarini, C. Baccarani, and V. Mascherpa, "Lean production, Toyota production system and Kaizen philosophy," *The TQM Journal*, vol. 30, no. 4, pp. 425–438, Jul. 2018. doi: 10.1108/TQM-12-2017-0178.
25. B. J. Galli, K. Ramirez, and E. Garcia, "Understanding the impact of quantitative decision-making on engineering managers," *IEEE Engineering Management Review*, vol. 51, no. 4, pp. 10–14, Dec. 2023. doi: 10.1109/EMR.2023.3292235.
26. V. Kumar, P. Verma, A. Mittal, P. Gupta, R. Raj, and M. S. Kaswan, "Addressing The kaizen business operations: The role of triple helix actors during COVID-19 outbreak," *The TQM Journal*, vol. 36, no. 6, pp. 1665–1694, Jun. 2024. doi: 10.1108/TQM-08-2023-0253.
27. T. Ohno, and N. Bodek, *Toyota Production System*. Productivity Press, 2019. doi: 10.4324/9780429273018.
28. M. da S. B. Gama, and A. Bonamigo, "Sustainable lean manufacturing as long-term strategy: Performance framework development and prioritization," *Journal of Strategy and Management*, vol. 17, no. 2, pp. 205–221, May 2024. doi: 10.1108/JSMA-05-2023-0104.
29. E. Psomas, and J. Antony, "Research gaps in Lean manufacturing: A systematic literature review," *International Journal of Quality & Reliability Management*, vol. 36, no. 5, pp. 815–839, May 2019. doi: 10.1108/IJQRM-12-2017-0260.
30. M. Núñez-Merino, J. M. Maqueira-Marín, J. Moyano-Fuentes, and P. J. Martínez-Jurado, "Information and digital technologies of Industry 4.0 and Lean supply chain management: A systematic literature review," *International Journal of Production Research*, vol. 58, no. 16, pp. 5034–5061, Aug. 2020. doi: 10.1080/00207543.2020.1743896.
31. Shingo Institute, "Shingo Model booklet v15.1," vol. 1, Jun. 2022, Accessed: Jun. 15, 2024. [Online]. Available: https://shingo.org/shingo-model/

32. R. G. Cooper, and P. Fürst, “Agile development in manufacturing companies: Best Practices and pitfalls,” *IEEE Engineering Management Review*, vol. 51, no. 4, pp. 65–76, Dec. 2023. doi: 10.1109/EMR.2023.3304792.
33. E. Nedjwa, R. Bertrand, and S. Sassi Boudemagh, “Impacts of Industry 4.0 technologies on Lean management tools: A bibliometric analysis,” *International Journal on Interactive Design and Manufacturing (IJIDeM)*, vol. 16, no. 1, pp. 135–150, Mar. 2022. doi: 10.1007/s12008-021-00795-9.
34. A. Bonamigo, R. R. de Azeredo, J. E. Monteiro de Camargo Filho, and H. de Souza Andrade, “Lean 4.0 in the value co-creation in agro-industrial services: An agenda for future studies for the efficient resource use,” *Systems Research and Behavioral Science*, vol. 41, no. 2, pp. 219–234, Mar. 2024. doi: 10.1002/sres.2950.

9 Lean Service 5.0

Drivers Analyses to Handling the Challenges of Changing Organizational Culture in B2B Relationships

Andrei Bonamigo, Keith Heringer dos Santos, Steffan Macali Werner, and José Carlos Sá

9.1 INTRODUCTION

Lean Thinking, when implemented effectively, results in the reduction of internal variability and uncertainty in the external environment, improving production quality, as well as reducing costs, contributing to increased operational performance (Dinis-Carvalho et al., 2023; Torre and Bonamigo, 2024; However, for the lean approach to be effective, the people who make up this system must be able to recognize the complex problems of a modern industrial system and have the ability to define and implement corrective and proactive actions. All employees must be fully aligned with the organizational culture (Vukadinovic et al., 2019).

The inhibitors to the implementation of the lean approach in the services sector have a strong relationship with individual characteristics of commitment, involvement, and preparation of personnel in an organizational environment. Therefore, for the improvement of a system to be effective, practices need to be human-based, including actions such as training and stimulating practices for integrating employees into the organization's culture (Ferenhof et al., 2018). Organizational culture, within the Lean approach, has three fundamental pillars: Respect for people, Continuous Improvement, and Management Aspects (Werner, 2021). For people to get involved in implementing the Lean approach, there must be change management to facilitate the transition of individuals and the development of organizational culture (AlManei et al., 2018). For that matter, in the context of Business-to-Business (B2B) value co-creation, in which there are interactions between multiple organizations with different skills, strategies, and resource complementarities (Bonamigo et al., 2020), management culture is fundamental so that there is a harmonious relationship, so that the compatibility of culture and business philosophy are integrated for co-creation to occur effectively (Steers et al., 2023). Since an important characteristic of I5.0 is the vision of human centralization at the heart of organizational processes, the aim is to employ technologies to work in favor of the employee, in symbiosis,

DOI: 10.1201/9781003535188-9

not with a function of replacing the human in the process (Bonamigo et al., 2024; Kaswan et al., 2024; Kumar et al., 2024a).

The lean service 5.0, in turn, "creates a learning environment to promote a lean mindset, share advance knowledge and orchestrate I5.0 technology" (Bonamigo et al., 2024; Sharma et al., 2023). Since companies focus on technology and not people, it is necessary to understand how technologies within Lean Service 5.0 can help culture facilitate B2B co-creation processes. According to Bonamigo et al. (2024), the combination of the lean approach and I5.0 in service organizations leads to LS 5.0. The concept is "defined as the integration and symbiosis of I5.0 technologies and principles with the lean services aiming to create, enhance, and deliver customer value through a socio-technical system based on technology" (Kumar et al., 2024b). Since, it allows providing a learning environment to promote a lean mindset, share and advance knowledge in addition to orchestrating I5.0 technology.

Based on the above, this chapter aims to present the drivers for cultural change in B2B value co-creation processes from the perspective of Lean Service 5.0. To guide the achievement of this objective, the following research questions were performed: *RQ1. What are the drivers to guide cultural change in co-creative environments from the perspective of LS5.0? RQ2. What is the impact of drivers for cultural change based on LS 5.0 in the context of value co-creation in services?*

9.2 METHOD

A systematic literature review (SRL) was conducted based on Ferenhof and Fernandes (2016). Then, the content was analyzed as recommended by Bardin (2011), the analysis was split into three phases: (i) Pre-analysis; (ii) exploration material or coding; and (iii) result's treatment, inference, and interpretation.

To accomplish the objective, we guided our literature review by the method of Systematic Search Flow (SSF) proposed by Ferenhof and Fernandes (2016). SSF Method is composed of four phases and eight activities. First, we established a research plan embracing the research questions we were interested in answering, which also involved the keywords and a set of inclusion and exclusion criteria. The research questions are the following: (i) *What are the key elements to guide cultural change in co-creative environments from the perspective of LS 5.0?* (ii) *What is the impact of drivers for cultural change based on LS 5.0 in the context of value co-creation in services?*

Then, we defined the research query *((Lean AND Service) AND (("industry 5.0" OR "I5.0" OR "human-centred" OR "human-centered" OR ("socio-technical" AND (Technological OR Technology))*. The inclusion criteria were empirical papers, peer-reviewed, English, Spanish, Portuguese, and indexed in the databases: Compendex, Ebesco, Scopus, Emerald, Science Direct, ProQuest, and WoS.

The exclusion criteria were the elimination of gray literature such as reports, conference papers, and non-academic research, besides in other languages. We developed a "knowledge matrix" as recommended by the SSF Method in a data sheet consisting of relevant criteria for establishing our understanding of LS 5.0 and the scope of this research about LS 5.0 and organizational culture.

Second, one among us entered the databases and queried them employing a combination of predefined keywords, looking for this combination in the articles' titles, keywords, and abstracts. The databases were searched on July 9, 2023, yielding 1787 documents. Fourth, each of us reviewed the abstracts of the respective documents. If relevant, we read parts of the full text to ensure it aligned with our scope of interest. This filtering process narrowed it down to 1477 documents meeting our criteria.

Fifth, we created a bibliographic portfolio for analysis. We exported the author (year), title, and journal information from the reference management software to a spreadsheet. We carefully reviewed each entry, editing the content and coding it based on the criteria outlined in Table 9.1.

In the Synthesis stage, we consolidated the data from each author into a single spreadsheet. This review enabled us the organize the findings into themes, shedding light on insights related to drivers, key factors, inhibitors, barriers, wastes, and advantages of LS 5.0 in services, co-creation, and organizational culture. Collaboratively, all authors participated in identifying themes for each entry, engaging in discussion rounds that also resulted in further paper reductions. Moving on to the eighth stage, the final phase of our review process focused on documenting and presenting the findings.

The content analysis comprised three main stages, following the approach recommended by Bardin (2011): (i) Pre-analysis; (ii) Examination of material through coding; and (iii) Treatment of results, inference, and interpretation. The process unfolded as outlined below.

During this phase, as part of the paper portfolio, authors examined the references of the initial 70 documents. This process detected an additional 11 works that were pertinent to the theme and subsequently incorporated into the final bibliographic portfolio. Following this, the 11 newly identified documents were coded and analyzed according to the content analysis criteria (Bardin, 2011).

Subsequently, individual data from the analyzed papers were synthesized into a consolidated spreadsheet. Each team member then independently reviewed the

TABLE 9.1
Resulting Bibliographic Portfólio

Database	Number of Documents
ProQuest	1300
Science Direct	1246
Emerald	657
Compendex	21
Scopus	20
ISI Web of Science	15
Ebesco	5
Subtotal	3264
Duplicates	(−1787)
Total	**1477**

Source: The Authors.

merged sheet to ensure consistency in coding regarding the context unit and record unit. Ultimately, the analysis was based on 18 papers, as detailed in Table 9.2. Subsequently, the conclusive phase of our review process centered on articulating the findings. The united analysis was defined as "*a posteriori*".

9.3 RESULTS AND DISCUSSION

From the content analysis based on the methodology exposed in session 3, 18 Record Units were extracted to support the drivers of organizational cultural change in B2B relationships as shown in Table 9.2.

TABLE 9.2
Documents Addressing Organizational Culture Drivers in B2B Relationships

Code	Authors (Year)	Title
A1	Villalba-Diez and Ordieres-Meré (2021)	Human-machine integration in processes within Industry 4.0 management
A2	Voulgaridis et al. (2022)	IoT and digital circular economy: Principles, applications, and challenges
A3	Da Silva and Marques (2020)	Human-centered design for collaborative innovation in knowledge-based economies
A4	AlManei et al. (2017)	Lean implementation frameworks: the challenges for SMEs
A5	Chuang (2024)	Indispensable skills for human employees in the age of robots and AI
A6	Naveed et al. (2022)	Do organizations really evolve? The critical link between organizational culture and organizational innovation toward organizational effectiveness: pivotal role of organizational resistance
A7	Vukadinovic et al. (2019)	Early management of human factors in lean industrial systems
A8	Basu and Dan (2020)	A comprehensive study of manifests in lean manufacturing implementation and framing an administering model
A9	AlManei et al. (2018)	A conceptual lean implementation framework based on change management theory
A10	Knapić et al. (2023)	Importance of first-line employees in lean implementation in SMEs: a systematic literature review
A11	Bakke and Johansen (2019)	Implementing of Lean – challenges and lessons learned
A12	Sulistyaningsih (2023)	Improving human resources technology innovation as a business growth driver in the society 5.0 era
A13	Valentinetti and Muñoz (2021)	Internet of things: emerging impacts on digital reporting
A14	Rot and Sobinska (2018)	The potential of the internet of things in knowledge management system
A15	van Dun and Kumar (2023)	Social enablers of Industry 4.0 technology adoption: transformational leadership and emotional intelligence
A16	Senadjki et al. (2023)	Unlocking the potential: the impact of digital leadership on firms' performance through digital transformation
A17	Gilli et al. (2024)	The future of leadership: new digital skills or old analog virtues?
A18	Bagga et al. (2023)	The mediating role of organizational culture: transformational leadership and change management in virtual teams

Based on the investigation, four drivers for organizational cultural change in companies with a B2B relationship were identified. They are Leadership Commitment, Training, Knowledge Sharing, and Communication. Table 9.3 depicts organizational culture drivers in B2B relationships

Leadership support is critical to the success of the change process (Bakke and Johansen, 2019). ISO 9001:2015 informs that leadership has a fundamental role in ensuring the implementation of cultural change with quality (Walaszczyk and Polak-Sopinska, 2020). Managers therefore need to act responsibly, encouraging teamwork and encouraging them to provide feedback to cultivate autonomy and motivation (Knapić et al., 2023). Managers must present the reason for the change, and the benefits that employees will have from it, communicate with pride the vision of the

TABLE 9.3
Organizational Culture Drivers in B2B Relationships

Record Unit	Context Unit	Freq
Leadership Commitment	A4 – Top management needs to show commitment and involvement. A7 – Managers need to have a clear vision of the mission to be achieved. A8 – Management must show availability and establish priorities. A9 – Leadership and commitment are key factors for success. A10 – Management is considered a facilitator for changing organizational culture. A11 – Management commitment is vital to successful implementation. A15 – Managers need to pay attention to the emotions and insecurities that will arise in employees during the change process, providing training to increase employees' confidence. It is also necessary that the recruitment of new employees is carried out considering, in addition to technical capacity, the emotional capacity to adapt. A16 – Digital transformation is positively impacted by leaders with digital capabilities, as these leaders not only innovate themselves but also have digital knowledge and help their employees to be more digitally competent. A17 – Leaders need to establish a relationship through empathy and authenticity and maintain transparent communication. Furthermore, they need to act as role models when it comes to encouraging changes and adaptations to new technologies within the organization. A18 – A cultural change is only successful when there is good leadership involved.	10
Training	A4 – Knowledge management within the organization and developing employee skills is vital for implementation to be successful. A5 – HR professionals need to stimulate employees' human skills and make skills improvement more viable and flexible for workers. A7 – Training new managers is necessary to adapt to changes. A11 – Training for managers and employees will strengthen the legitimacy of Lean and cultural change. A12 – Employee training is essential for them to develop their potential; each organization needs to implement training to improve the quality of their employees' work.	5

(Continued)

TABLE 9.3 *(Continued)*
Organizational Culture Drivers in B2B Relationships

Record Unit	Context Unit	Freq
Knowledge Sharing	A1 – The use of IIOT can increase the transparency of processes, as well as the effectiveness of management decisions made. Technology provides an agnostic perspective on the process and its impact. A2 – The combination of the concept of Circular Economy and IoT is a facilitator as it decentralizes information and further integrates interested parties within the flow of a product/process through data sharing. A13 – IoT can be used to store information about stakeholders' behaviors and preferences, produce the information, and automatically address it to the users who need it. A14 – IoT can provide tools that enable access to technological and organizational knowledge, providing support for knowledge transfer and sharing.	4
Communication	A4 – Communication is considered a success factor: Communication channels between senior management and employees need to be active, and communication of change initiatives must be constant. A6 – Communication is important because employees will be resistant to change if they do not feel part of the process. A10 – Communication and teamwork are considered a facilitator for culture change. A18 – For an organization to respond positively to cultural changes, it is necessary to have strong communication and integration between systems.	4

future work environment (Liu et al., 2022), motivate, and invite employees to provide input to prepare for the implementation of the process, and offering personalized opportunities for employees to raise their potential concerns (van Dun and Kumar, 2023). For employees' skills to be continually developed, human resource development professionals must understand the essential competencies to be developed to direct everyone (Park et al., 2018), including the management that directs their subordinates (Chuang, 2024; Vukadinovic et al., 2019). Training is a development strategy for an organization (Tūtlys and Spöttl, 2022) because, through it, employees can develop skills that will later be applied in the organization (Sulistyaningsih, 2023; Susomrith et al., 2019).

Training is necessary to manage the organization's knowledge and improve employees' skills so that there is no asymmetry between organizations during the co-creation process (AlManei et al., 2017; Bonamigo et al., 2022; Bonamigo et al., 2024; Ferenhof et al., 2024). Furthermore, training contributes to continuous improvement as it brings legitimacy to the purpose of co-creating value (Bakke and Johansen, 2019). The lack of training becomes a barrier and brings risk to organizational changes (Hassan et al., 2024) as employees do not understand the purpose of their role in the face of the changes observed, and as a result, teamwork is compromised and increases conflicts between sectors (Sult et al., 2024; Torre and

Bonamigo, 2024). Disseminating knowledge within the organization is a prerequisite for applying it effectively in problem-solving and decision-making (Jarrahi et al., 2023). The exchange of information and complementarity of resources between organizations are necessary to obtain the benefits of value co-creation (Ferenhof et al., 2022; Ferenhof et al., 2024). In this context, Artificial Intelligence can be useful as it makes it possible to create more connected coordination systems, which give managers a better idea of knowledge constraints (Jarrahi et al., 2023).

Ineffective communication can result in a misunderstanding of the purpose and benefits of relationships between organizations (Bagga et al., 2023; Basu and Dan, 2020). As a result, employees may feel devalued and unwell, making the collaborative and integrated process difficult (Bakke and Johansen, 2019; Knapić et al., 2023). To avoid resistance from employees, it is necessary to cultivate engagement and a culture of teamwork (Chuang, 2024; Martinez et al., 2023). The use of collaborative digital tools, with the creation of virtual teams, reduces communication barriers and opens space for employees to bring their questions and ideas, therefore enriching relationships by increasing trust between those involved (Batra et al., 2024; Sabale et al., 2024; Stray and Moe, 2020).

To deal with the cultural change required in a B2B relationship, four drivers were identified. These drivers influence positively using the support of Industry 5.0 technologies together with Lean Thinking, to guide culture change through the centralization of human needs in organizational decisions. The relationship between these drivers and the co-creation players is shown in Figure 9.1.

According to Figure 9.1, we can highlight the elements found in this research as elements that allow guiding/facilitating the process of value co-creation from the perspective of cultural change from the perspective of LS.50. Leadership commitment contributes to employee engagement. Using capabilities such as big data for

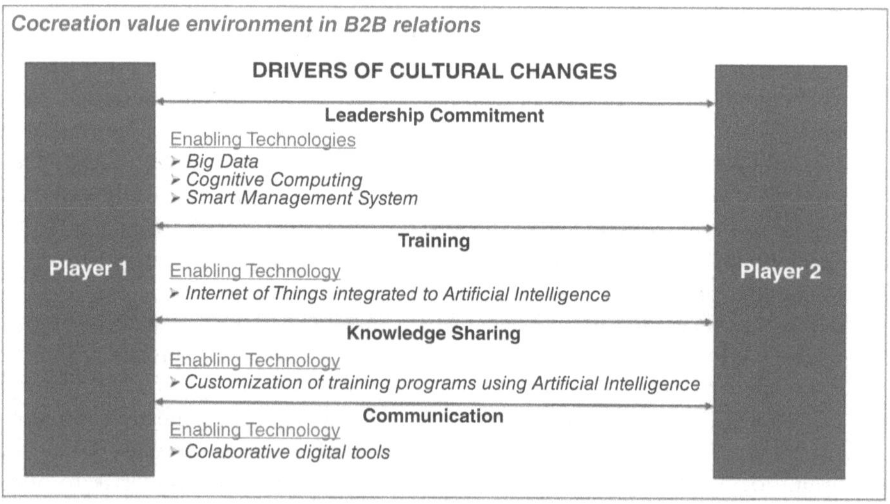

FIGURE 9.1 Organizational culture change drivers in B2B relationships. (The authors (2024).)

data collection and processing, Cognitive Computing to support decisions and the Smart Management system for efficient task allocation, leadership will be supported by enabling technologies, which favors integration between the companies. In turn, Training contributes to leveling and developing the skills of both management and employees, to acquire autonomy and confidence, in addition to understanding the motivation for cultural changes, to avoid resistance. Thus, Training favors the co-creation process by mitigating the knowledge gap, which is an inhibitor to the co-creation of value.

9.4 CONCLUSION

This research aimed to present the drivers for cultural change in B2B value co-creation processes from the perspective of Lean Service 5.0. To guide the culture change, we identified four key points: (i) Leadership Commitment, (ii) Training, (iii) Knowledge Sharing, and (iv) Communication. We can observe that the Leadership Commitment element is the one that represents the highest frequency, which indicates that this element is one of the main factors in ensuring the success of a change in organizational culture. Next, we observed that training mitigates the adverse factors of resistance to change and access to new knowledge and skills based on industry 5.0 technologies and from the perspective of Lean Thinking. In this way, people's engagement demands a relationship of trust to exchange information and share resources efficiently. The use of the IoT minimizes time and effort as it optimizes the management of information obtained in this interaction and, together with analysis tools, contributes to decision-making that could be difficult due to cultural conflicts between actors who jointly co-create value. For future studies, it is possible to analyze the impact and ways of implementing each driver isolated and evaluate whether other industry 5.0 technologies can be used as facilitators.

REFERENCES

AlManei, M., Salonitis, K. and Tsinopoulos, C. (2018), "A conceptual lean implementation framework based on change management theory", Procedia CIRP, Vol. 72, pp. 1160–1165. https://doi.org/10.1016/j.procir.2018.03.141

AlManei, M., Salonitis, K. and Xu, Y. (2017), "Lean implementation frameworks: the challenges for SMEs", Procedia CIRP, Vol. 63, pp. 750–755. https://doi.org/10.1016/j.procir.2017.03.170

Bagga, S.K., Gera, S. and Haque, S.N. (2023), "The mediating role of organizational culture: transformational leadership and change management in virtual teams", Asia Pacific Management Review, Vol. 28 No. 2, pp. 120–131. https://doi.org/10.1016/j.apmrv.2022.07.003

Bakke, A.L. and Johansen, A. (2019), "Implementing of Lean–challenges and lessons learned", Procedia Computer Science, Vol. 164, pp. 373–380. https://doi.org/10.1016/j.procs.2019.12.196

Bardin, L. (2011), *Análise de Conteúdo, 3. reimp*, Lisboa.

Basu, P. and Dan, P.K. (2020), "A comprehensive study of manifests in lean manufacturing implementation and framing an administering model", International Journal of Lean Six Sigma, Vol. 11 No. 4, pp. 797–820. https://doi.org/10.1108/IJLSS-11-2017-0131

Batra, I., Sharma, C., Malik, A., Sharma, S., Kaswan, M.S. and Garza-Reyes, J.A. (2024), "Industrial revolution and smart farming: a critical analysis of research components in industry 4.0", The TQM Journal, Vol. ahead-of-print No. ahead-of-print. https://doi.org/10.1108/TQM-10-2023-0317

Bonamigo, A., Dettmann, B., Frech, C.G. and Werner, S.M. (2020), "Facilitators and inhibitors of value co-creation in the industrial services environment", Journal of Service Theory and Practice, Vol. 30 No. 6, pp. 609–642. https://doi.org/10.1108/JSTP-03-2020-0061

Bonamigo, A., Frech, C.G. and Lopes, A.C.C. (2022), "Value co-creation in the B2B context: a diagnosis of knowledge management based on multiple case studies", Journal of Business & Industrial Marketing, Vol. 37 No. 7, pp. 1449–1462. https://doi.org/10.1108/JBIM-11-2020-0528

Bonamigo, A., Werner, S.M. and de Oliveira, M.S. (2024), "Toward lean service 5.0: evaluation of future perspectives and trends", Revista Meta: Avaliação, Vol. 16 No. 50, pp. 180–210. http://dx.doi.org/10.22347/2175-2753v16i50.4335

Chuang, S. (2024), "Indispensable skills for human employees in the age of robots and AI", European Journal of Training and Development, Vol. 48 No. 1/2, pp. 179–195. https://doi.org/10.1108/EJTD-06-2022-0062

Da Silva, T.F.P. and Marques, T.F.P. (2020), "Human-centered design for collaborative innovation in knowledge-based economies", Technology Innovation Management Review, Vol. 10 No. 9, pp. 5–15. http://dx.doi.org/10.22215/timreview/1385

Dinis-Carvalho, J., Sousa, R.M., Moniz, I., Macedo, H. and Lima, R.M. (2023), "Improving the performance of a SME in the cutlery sector using lean thinking and digital transformation", Sustainability, Vol. 15 No. 10, p. 8302. https://doi.org/10.3390/su15108302

Ferenhof, H.A., Bonamigo, A., Rosa, L.G. and Vieira, T.C. (2024), "Theoretical B2B knowledge management framework focused on value co-creation", VINE Journal of Information and Knowledge Management Systems, Vol. 54 No. 2, pp. 424–451. https://doi.org/10.1108/VJIKMS-10-2021-0239

Ferenhof, H.A., Da Cunha, A.H., Bonamigo, A. and Forcellini, F.A. (2018), "Toyota Kata as a KM solution to the inhibitors of implementing lean service in service companies", VINE Journal of Information and Knowledge Management Systems, Vol. 48 No. 3, pp. 404–426. https://doi.org/10.1108/VJIKMS-11-2017-0083

Ferenhof, H.A. and Fernandes, R.F. (2016), "Demystifying the literature review as basis for scientific writing: SSF method", Revista ACB, Vol. 21 No. 3, pp. 550–563.

Gilli, K., Lettner, N. and Guettel, W. (2024), "The future of leadership: new digital skills or old analog virtues?, Journal of Business Strategy, Vol. 45 No. 1, pp. 10–16. https://doi.org/10.1108/JBS-06-2022-0093https://doi.org/10.1108/JBS-06-2022-0093

Hassan, M.A., Zardari, S., Farooq, M.U., Alansari, M.M. and Nagro, S.A. (2024), "Systematic analysis of risks in industry 5.0 architecture", Applied Sciences, Vol. 14 No. 4, p. 1466. https://doi.org/10.3390/app14041466

Jarrahi, M.H., Askay, D., Eshraghi, A. and Smith, P. (2023), "Artificial intelligence and knowledge management: a partnership between human and AI", Business Horizons, Vol. 66 No. 1, pp. 87–99. https://doi.org/10.1016/j.bushor.2022.03.002

Kaswan, M.S., Chaudhary, R., Garza-Reyes, J.A. and Singh, A. (2024), "A review of industry 5.0: from key facets to a conceptual implementation framework", International Journal of Quality & Reliability Management, Vol. ahead-of-print No. ahead-of-print. https://doi.org/10.1108/IJQRM-01-2024-0030

Knapić, V., Rusjan, B. and Božič, K. (2023), "Importance of first-line employees in lean implementation in SMEs: a systematic literature review", International Journal of Lean Six Sigma, Vol. 14 No. 2, pp. 277–308. https://doi.org/10.1108/IJLSS-08-2021-0141

Kumar, U., Kaswan, M.S., Kumar, R., Chaudhary, R., Garza-Reyes, J.A., Rathi, R. and Joshi, R. (2024a), "A systematic review of industry 5.0 from main aspects to the execution status", The TQM Journal, Vol. 36 No. 6, pp. 1526–1549. https://doi.org/10.1108/TQM-06-2023-0183

Kumar, V., Verma, P., Mittal, A., Gupta, P., Raj, R. and Kaswan, M.S. (2024b), "Addressing the Kaizen business operations: the role of triple helix actors during COVID-19 outbreak", The TQM Journal, Vol. 36 No. 6, pp. 1665–1694. https://doi.org/10.1108/TQM-08-2023-0253

Liu, S.F., Fan, Y.J., Luh, D.B. and Teng, P.S. (2022), "Organizational culture: the key to improving service management in industry 4.0", Applied Sciences, Vol. 12 No. 1, p. 437. https://doi.org/10.3390/app12010437

Martinez, N., Kilag, O.K. and Macario, R. (2023), "The impact of organizational culture on leadership strategies in crisis management", Excellencia: International Multi-Disciplinary Journal of Education (2994-9521), Vol. 1 No. 5, pp. 454–466. Available at: https://multijournals.org/index.php/excellencia-imje/article/view/145 (Accessed: 29 February 2024).

Naveed, R.T., Alhaidan, H., Al Halbusi, H. and Al-Swidi, A.K. (2022), "Do organizations really evolve? The critical link between organizational culture and organizational innovation toward organizational effectiveness: pivotal role of organizational resistance", Journal of Innovation & Knowledge, Vol. 7 No. 2, p. 100178. https://doi.org/10.1016/j.jik.2022.100178

Park, S., Kang, H.-S.T. and Kim, E.-J. (2018), "The role of supervisor support on employees' training and job performance: an empirical study", European Journal of Training and Development, Vol. 42 No. 1/2, pp. 57–74. https://doi.org/10.1108/EJTD-06-2017-0054

Rot, A. and Sobinska, M. (2018). The potential of the internet of things in knowledge management system. In FedCSIS (Position Papers) (pp. 63–68).

Sabale, D.B., Kaswan, M.S., and Rathi, R. (2024), Development of circular economy model using Industry 4.0's machine learning techniques to improve performance metrics of MSMEs, International Journal on Interactive Design and Manufacturing (IJIDeM), 1–19.

Senadjki, A., Au Yong, H.N., Ganapathy, T. and Ogbeibu, S. (2023), "Unlocking the potential: the impact of digital leadership on firms' performance through digital transformation", Journal of Business and Socio-Economic Development. https://doi.org/10.1108/JBSED-06-2023-0050

Sharma, S., Malik, A., Sharma, C., Batra, I., Kaswan, M.S. and Garza-Reyes, J.A. (2023), "Adoption of industry 4.0 in different sectors: a structural review using natural language processing", International Journal on Interactive Design and Manufacturing (IJIDeM), pp. 1–23. https://doi.org/10.1007/s12008-023-01550-y

Steers, R.M., Osland, J.S. and Szkudlarek, B. (2023), *Management Across Cultures: Challenges, Strategies, and Skills*, Cambridge University Press.

Stray, V. and Moe, N.B. (2020), "Understanding coordination in global software engineering: A mixed-methods study on the use of meetings and Slack", Journal of Systems and Software, Vol. 170, p. 110717. https://doi.org/10.1016/j.jss.2020.110717

Sulistyaningsih, E. (2023), "Improving human resources technology innovation as a business growth driver in the society 5.0 era", ADI Journal on Recent Innovation, Vol. 4 No. 2, pp. 149–159. https://doi.org/10.34306/ajri.v4i2.820

Sult, A., Wobst, J. and Lueg, R. (2024), "The role of training in implementing corporate sustainability: a systematic literature review", Corporate Social Responsibility and Environmental Management, Vol. 31 No. 1, pp. 1–30. https://doi.org/10.1002/csr.2560

Susomrith, P., Coetzer, A. and Ampofo, E. (2019), "Training and development in small professional services firms", European Journal of Training and Development, Vol. 43 No. 5/6, pp. 517–535. https://doi.org/10.1108/EJTD-11-2018-0113

Torre, N. and Bonamigo, A. (2024), "Action research of lean 4.0 application to the maintenance of hydraulic systems in steel industry", Journal of Quality in Maintenance Engineering, Vol. 30 No. 2, pp. 341–366. https://doi.org/10.1108/JQME-06-2023-0058

Tūtlys, V. and Spöttl, G. (2022), "Disruption of qualifications in manufacturing: challenges and prospects", European Journal of Training and Development, Vol. 46 No. 5/6, pp. 390–412. https://doi.org/10.1108/EJTD-07-2020-0121

Valentinetti, D. and Muñoz, F.F. (2021), "Internet of things: emerging impacts on digital reporting", Journal of Business Research, Vol. 131, pp. 549–562. https://doi.org/10.1016/j.jbusres.2021.01.056

van Dun, D.H. and Kumar, M. (2023), "Social enablers of industry 4.0 technology adoption: transformational leadership and emotional intelligence", International Journal of Operations & Production Management, Vol. 43 No. 13, pp. 152–182. https://doi.org/10.1108/IJOPM-06-2022-0370

Villalba-Diez, J. and Ordieres-Meré, J. (2021), "Human–machine integration in processes within industry 4.0 management", Sensors, Vol. 21 No. 17, p. 5928. https://doi.org/10.3390/s21175928

Voulgaridis, K., Lagkas, T., Angelopoulos, C.M. and Nikoletseas, S.E. (2022), "IoT and digital circular economy: principles, applications, and challenges", Computer Networks, Vol. 219, p. 109456. https://doi.org/10.1016/j.comnet.2022.109456

Vukadinovic, S., Macuzic, I., Djapan, M. and Milosevic, M. (2019), "Early management of human factors in lean industrial systems", Safety Science, Vol. 119, pp. 392–398. https://doi.org/10.1016/j.ssci.2018.10.008

Walaszczyk, A. and Polak-Sopinska, A. (2020). "The role of leadership in organizations managed in conformity with ISO 9001 quality management system standard". In: *Kantola, J., Nazir, S. (eds) Advances in Human Factors, Business Management and Leadership. AHFE 2019. Advances in Intelligent Systems and Computing*, vol 961. Springer, Cham. https://doi.org/10.1007/978-3-030-20154-8_37

Werner, S.M. (2021). "Modelo de aprendizagem baseada em projetos para implementação de lean baseado em ciclos curtos de experimentação". *PhD Thesis*, Universidade Federal de Santa Catarina, Brazil.

10 Industry 5.0 and the Workers' Engagement

A Framework Based on the Toyota Kata Approach to Guide Operational Excellence

Raphael Odebrecht de Souza, Andrei Bonamigo, and Fernando Antônio Forcellini

10.1 INTRODUCTION

The incorporation of new digital technologies into manufacturing systems gave rise to the term Industry 4.0 (I4.0) (Lu, 2017; Torre, Bonamigo, 2024; Sabale et al., 2024). Considering a way of increasing productivity, the objective of I4.0 is to interconnect processes and simulate optimizations on the Cyber-Physical System (CPS) before implementing the real environment (Tortorella, Fettermann, 2018). Although I4.0 has brought several benefits to industries in different sectors, such as manufacturing, agriculture, and energy (Sharma et al., 2023; Kumar et al., 2024a), the role of workers was not fully addressed in this industrial revolution. Thus, after the maturation of these concepts in the past decade, recent research stated the term Industry 5.0 (I5.0) as a new way to bring workers to the center of the discussion (Kumar et al., 2024b).

In contrast to I4.0, the main goal of I5.0 is to harness the creative potential of humans and the technological capabilities of new technologies to achieve lasting solutions that are more efficient (Maddikunta et al., 2022). In this context, I5.0 is a human-driven concept aiming to bring workers back to the center of the discussion, as opposed to I4.0 which is technologically driven (Batra et al., 2024; Leng et al., 2022). Kaswan et al. (2024) suggest that I5.0 concepts are still being created. Thus, new theoretical frameworks are needed to better understand its evolution and applicability.

One strategy that stands out in this human-centric discussion is the Toyota Kata (TK) approach. The approach described by Rother in 2010 is used by Toyota to continually improve its processes with the involvement of workers. It is a routine based on the scientific method that aims to develop new strategies or tools that reduce waste and improve efficiency. Given the human-centric nature of the TK approach, recent studies, such as those from Odebrecht de Souza et al. (2022), Eren (2024),

DOI: 10.1201/9781003535188-10

Ríos-Hernández et al. (2024), propose incorporating TK and I4.0 concepts to facilitate the integration of digital technologies in industries.

10.2 LITERATURE REVIEW

This section presents a brief literature review of key topics used as constructs for the proposed framework.

The term "Lean" emerged in the late 1980s from a Massachusetts Institute of Technology (MIT) study of the global auto manufacturing industry, which highlighted Toyota's innovative business management model across different dimensions. Although Lean originated in the automotive sector, it gained widespread application in other areas following the release of the book "Lean Thinking" by Womack and Jones in 1996. The book defined the core principles of Lean: identify value from the customer perspective, map the value stream, create flow by reducing waste, establish pull, and seek perfection through continued improvements. Ohno (1988) identified seven basic wastes in production environments that should be minimized: inventory, waiting, defects, overproduction, motion, transportation, and over-processing.

To visualize product and information flow in a process, from the entry of raw material to finished products, Lean implementation is usually accompanied by an approach called Value Stream Mapping (VSM) (Rother, Shook, 1998). The VSM approach consists of four steps: (1) preparation, (2) current state map, (3) future map, and (4) planning and implementation. Lean implementation in organizations with the full involvement of people is usually complex and shouldn't be taken as a project but as a journey (Houshmand and Jamshidnezhad, 2006). Considering that the implementation of Lean is an interrelated system between workers and tools, the vast majority of the organizations focus their efforts only on the use of the classical Lean tools developed by Toyota, without considering the change in organizational culture. In this way, they are rarely able to produce lasting changes or improvements (Bortolotti et al., 2015). Rother (2010) explored Toyota's methods for sustaining continuous improvement and introduced an approach within the company that he termed TK. Kata refers to a routine or "way of doing" and is characterized as a pattern that is repeated continuously, developing skills and a new mindset for continuous improvement in workers. This pattern directly interferes with the way employees think and behave, as it is a routine of continuous improvement that provides training for people based on the application of the scientific method for way-sustainably solving problems (de Souza et al., 2023b). The TK approach aims to incorporate continuous improvement thinking in employees (Reverol, 2012). According to Rother (2010), in a process improvement situation, there is an unknown path between the desired performance (Target Condition) and the current state of the process (Current Condition). Continuous improvement is the capacity to navigate toward a desired condition through uncertain terrain, remaining responsive to it, and adapting to the changing conditions.

10.3 METHOD

In line with the research objectives, a Systematic Literature Review (SLR) research was conducted using the PRISMA method from Moher et al. (2009). The content analysis methodology followed the method for reviewing papers described by

TABLE 10.1
Papers in Line with Research Objectives

Authors	Title
Ríos-Hernández et al. (2024)	Emotional intelligence and analytical skill as human capital factors influencing competitiveness through analysis of Toyota Kata, Kaizen, and Quality 4.0: an empirical study conducted during COVID-19.
Frank et al. (2024)	Beyond Industry 4.0 – integrating Lean, digital technologies, and people.
Rahardjo et al. (2024)	A sustainable innovation framework based on Lean Six Sigma and Industry 5.0.
Köhler et al. (2024)	Human-centered optimization of assembly systems with the help of design thinking.
Fani et al. (2024)	Lean and Industry 4.0 principles toward Industry 5.0: a conceptual framework and empirical insights from the fashion industry.
Faisal et al. (2024)	Development of a conceptual model for the introduction stage of lean intelligent manufacturing using a simulation model based on the identification of the appropriate technologies of Industry 5.0.
Eren (2024)	Toyota Kata as a Scaffolding for Human-Centric Manufacturing: Applying Lean Thinking for a Digital and Sustainable Factory of the Future.
Moraes et al. (2023)	Lean and Industry 4.0: A review of the relationship, its limitations, and the path ahead with Industry 5.0.

Source: The authors.

Watson and Webster (2020), and it involved three steps: (1) identifying the relevant literature; (2) structuring the review; and (3) theoretical development.

In (1) identifying the relevant literature, we used the database Scopus and defined the keywords and Boolean operators. Thus, the following query string was used in the searches: ((lean OR "Toyota Kata" AND "industr* 5.0") OR ("Toyota Kata" AND "Industr* 4.0")). Searches were conducted on 15 July 2024. In total, the search retrieved 59 articles. The PRISMA method was used to exclude studies that did not fall within the research scope. With the help of Endnote® software, 42 articles were excluded after reading the titles, keywords, and abstracts, as they were not related to the theme under research, and nine PDF files were not found. Finally, the research portfolio was composed of eight research papers, as shown in Table 10.1.

Following the research methodology, in structuring the review (2), data was retrieved from the articles to structure the review. In the step of theoretical development (3), the articles were screened and critically evaluated based on the TK approach, I4.0, and I5.0 concepts. Finally, using concepts found in the literature review, a framework to guide workers in their daily routine to implement digital technologies in manufacturing systems was proposed.

10.4 FINDINGS

Few studies have explored the integration of Lean methodologies, I4.0 technologies, and I5.0 concepts. Authors Ríos-Hernández et al. (2024) highlighted the influence of human capital factors like emotional intelligence and analytical skills on organizational competitiveness within the automotive parts industry, emphasizing the development of TK and Kaizen approaches for enhancing Quality 4.0. Authors Frank et al. (2024) have proposed a framework that integrates Lean and I4.0 that aims to improve workers' problem-solving capabilities. In order to make improvements more human-centered, Rahardjo et al. (2024) proposed a framework that combines Lean Six Sigma and I5.0 concepts. Similarly, Köhler et al. (2024) used the Design Thinking concept to optimize assembly processes. The theoretical and empirical results of these studies showed that human-centered strategies can promote the autonomy of employees to propose innovative solutions to specific problems.

Further studies by authors Fani et al. (2024), Faisal et al. (2024), Eren (2024), Moraes et al. (2023), and Bonamigo et al. (2024) also explored the synergies between Lean methodologies and emerging digital technologies in diverse industries scenarios. A Lean 5.0 framework for the fashion industry was proposed by Fani et al. (2024). The authors have combined Lean practices with I4.0 technologies, and the effectiveness of the framework was validated in an empirical case study. Faisal et al. (2024) achieved similar results with their own Lean and I5.0 framework. Authors Eren (2024) have developed a framework based on the TK approach. To the authors, the human-centered approach of TK is fundamental in the context of I5.0. The research of Moraes et al. (2023) and Bonamigo et al. (2024) highlighted the role of leadership and employee engagement as critical components in promoting a sustainable, resilient, and human-centered industrial environment in the context of Lean and I5.0.

Despite the advancements of the most recent research about the theme, a gap remains on how to effectively implement digital technologies in production processes with the participation of workers. While some authors have discussed the role of TK, no study presented a step by step to achieve the transition to I5.0. Thus, the next section presents the proposed framework aimed at filling this gap.

10.4.1 Proposed Framework

Given the discussion of the previous discussion, this section aims to present the proposed framework. The proposed framework integrates elements from established concepts while introducing new components when necessary. It does not aim to reinvent existing methodologies but rather to address gaps in current knowledge.

To conduct this study, we use the definition of Framework from Shehabuddeen et al. (2006) which states that: "A framework supports understanding and communication of structure and relationship within a system for a defined purpose". In this way, the system is the production system with several elements. The purpose is to implement the technologies from I4.0, but in a way that workers are elevated to smart workers who can use their creativity to solve problems through the TK approach. The framework has several layers, but the core is I5.0. The conduction of improvement is based on the Improvement Kata and Coaching Kata routine, resulting in the sequencing of target conditions.

As continuous improvement is driven by people, to this day, achieving it remains impossible without their involvement. In this way the TK approach favors the following factors: (1) process executors actively participate in activities and building

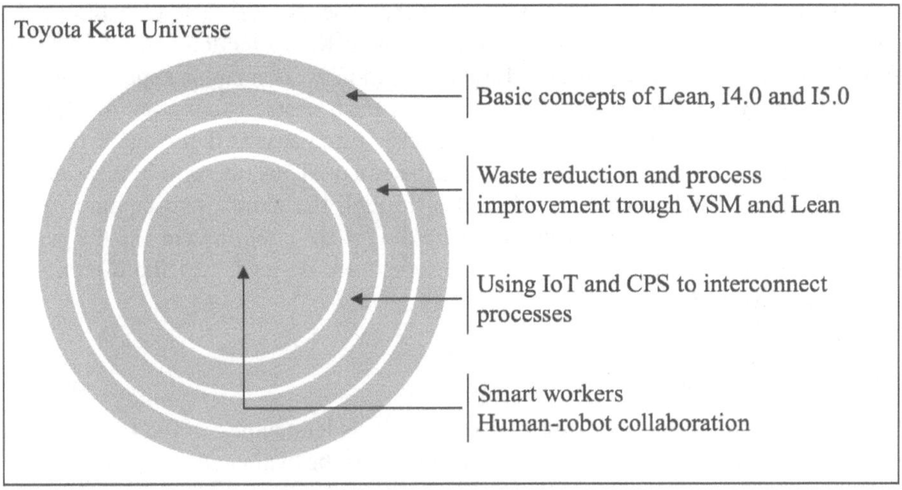

FIGURE 10.1 Layers of the proposed framework. (Source: The authors.)

improvements, (2) the improvements are in line with the organization's strategic objectivities, (3) fostering the ability to generate ideas and design innovative solutions, (4) the ability to reflect on each action planned and carried out, verifying the results obtained.

10.4.2 Proposed Framework Overview

Figure 10.1 illustrates the proposer framework layers. All the layers are embedded in the TK universe since the using scientific method throughout the PDCA cycle is crucial to conducting improvement actions (Rother, 2010).

The outermost layer includes the fundamental Lean, I4.0, and I5.0. These are the basic concepts that the improvement team must be familiar with. The next layer focuses on waste reduction using Lean principles and VSM, which helps visualize and identify waste while proposing future optimized states (Rother, Shook, 1998). Once the production process has been optimized and flow becomes continuous, IoT and CPS can be used to interconnect processes (de Souza et al., 2022). In this layer, the CPS can be used to simulate future states and improvements can be replicated in the digital environment before its implementation on the real system. At the core of the framework are the principles of I5.0, where I4.0 technologies in production processes support and empower employees. The goal is to use technological resources to develop Smart Workers and foster Human-Robot collaboration.

The proposed framework consists of six phases: (1) Team training, (2) Development of the company's vision, (3) Value Stream Mapping, (4) Implementation of I4.0 technologies at the process level, (5) Development of the CPS, and (6) Integration of humans and technologies. Each step will be discussed in detail in the following sections.

Each step at the organizational level has its gate. Improvement initiatives with the implementation of I4.0 technologies and active worker participation in transitioning to I5.0 are aligned with the TK approach at the process level. The Coaching Kata operates on a vertical axis, while the Improvement Kata functions on a horizontal axis. To overcome obstacles in each Target Condition, workers must experiment with short PDCA cycles (Plan, Do, Check, Act). For each obstacle, an action is planned and implemented, its results are verified (checked), and a reflection is made on what happened and what was learned with the experiment. Thus, the aim is to reach the Target Condition by overcoming obstacles through the team's participation using PDCA cycles. The obstacles can change according to the environment and the people involved. However, the conditions to be reached are defined, and the direction is clear to all involved.

10.4.3 Team Training

Initially, the worker's current state related to their understanding of the key concepts required to implement the proposed framework must be evaluated. Thus, knowledge gaps should be resolved through team training in the fundamentals of Lean methodology, the Kata routine, I4.0, and I5.0.

Participants are expected to be familiar with the five principles of Lean: (1) identify value, (2) map the value stream, (3) create flow, (4) establish pull, and (5) seek perfection (Womack, Jones, 1996). It is also crucial for everyone involved to understand the Improvement and Coaching Kata routine. Related to I4.0 concepts, workers should first be familiarized with emerging technologies such as CPS, the IoT, Big Data, and Cloud Computing. Considering that one of the benefits of using the new technologies is the measurement of production process parameters in real time, emphasis should be placed on the IoT technologies, as they are the sensors that will be in direct contact with the machines and equipment and will be responsible for transmitting the data to the CPS.

Another important aspect is that new technologies should only be considered to be implemented if they support workers, as the I5.0 aims not to replace humans with machines. Thus, emphasis should be on concepts such as Smart Workers, Human-Robot collaboration, and human-centered problem-solving approaches.

At the end of the first stage, participants in the training process are expected to be familiar with the concepts and terminologies necessary to use the proposed framework. However, it is not expected that all participants have all the answers, since learning is a procedural process throughout and the concepts become familiar over time. In this context, leadership commitment plays a key role in establishing a culture of continuous improvement and innovation, as it significantly impacts organizational performance and employee engagement (Mittal et al., 2023). Through consistent support and engagement, leaders can foster an environment where learning is encouraged, and improvement initiatives are sustained.

10.4.4 Development of the Company's Vision

To ensure the conditions for using the TK routine, the direction of the company must be clear and its unfolding to the lower levels must be carried out. To ensure the conditions for using the TK routine, the direction must be clear and its unfolding to the lower levels must be carried out. Thus, based on a bigger objective of the organization, challenges for each team are defined. The challenges of the different levels will be linked to the objectives of the organization, with the new Target Condition of one level being the challenge of the lower level.

At the end of this stage, the improvement team must be sure that every level of the company is in syntony and pointed in the same direction. This is important to ensure that no time is wasted on activities that do not aggregate value to the company's objective.

10.4.5 Value Stream Mapping

It is from the VSM that the opportunities to improve processes, identify and reduce waste, and implement the technologies from I4.0 are identified. The VSM approach also ensures that the improvement does not arise from an isolated initiative but is directed toward what the company is prioritizing. The process must be carried out with the direct participation of everyone involved in the process flow, such as supervisors and employees at the operational level. Thus, real problems can be identified, and the correct definition and implementation of improvements can be done.

As mentioned before, the team must be clear that new technologies must assist employees and not replace them. In this way, priority should be given to the implementation of sensors and IoT technologies so that they allow a real-time view of the Current Condition of the system and how the system behaves. From this, it is possible to create a digital version of the production process in the CPS, which, in turn, allows its visualization in real time. Thus, the implementation of improvements can be projected by workers in the digital environment before being implemented in the actual process. It is also important to automate repetitive tasks with the development of human robots (cobots). In this type of collaboration, robots get repetitive tasks while workers use their creativity to develop new solutions in a constantly changing environment.

The VSM approach in conjunction with CPS and IoT technologies is a living document that should be constantly updated and improved. It is only possible to improve something that can be measured and, in this way, past and present states can be compared to generate optimized future states.

10.4.6 Implementation of I4.0 Technologies at the Process Level

From the value flow map and the breakdown of activities at the different hierarchical levels of the organization, the team can select a specific process to start the implementation of digital technologies with the participation of shop floor workers at the process level. The breakdown of these activities is shown schematically in Figure 10.2 with the selection of a work cell. This process must be implemented by Loops in all systems and subsystems of the value flow map, one at a time.

The alignment of improvement activities with the TK approach can be seen. The Target Condition of a process becomes the implementation of I4.0 technologies, represented schematically in this example by a sensor. To go from the Current Condition to the Target Condition, and solve problems and obstacles, the Improvement Kata

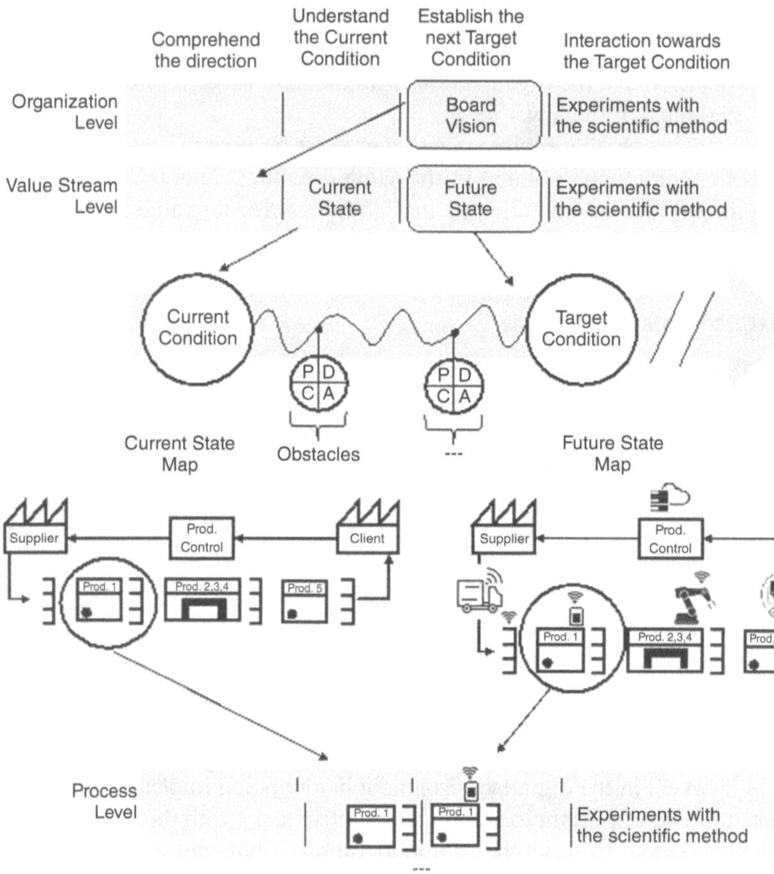

FIGURE 10.2 Unfolding of the company vision and TK on the processes level. (Source: The authors.)

should be carried out so that the challenge of implementing the technologies of I4.0 is achieved.

10.5 CONCLUSION, LIMITATIONS, AND FUTURE RESEARCH AGENDA

This research aimed to propose a framework based on TK to implement digital technologies in production processes as a human-centered approach. Based on existing gaps from the conducted SLR, the proposed framework aims to guide the daily routine of workers to continuously improve processes with the implementation of I4.0 technologies. Thus, with six steps and six gates, the proposed framework ensures that the workers are well trained on basic concepts, all organizational levels point in the same direction, the VSM is used to identify and reduce waste, I4.0 technologies can be implemented, the CPS is developed, and technologies are integrated with humans.

REFERENCES

Batra, I., Sharma, C., Malik, A., Sharma, S., Kaswan, M.S. and Garza-Reyes, J.A. (2024) 'Industrial revolution and smart farming: A critical analysis of research components in Industry 4.0', The TQM Journal (ahead-of-print). doi: 10.1108/TQM-10-2023-0317.

Bonamigo, A., Werner, S.M. and Oliveira, M.S. (2024) 'Toward lean service 5.0: Evaluation of future perspectives and trends', Revista Meta: Avaliação, 16(50), pp. 180–210. doi: 10.22347/2175-2753v16i50.4335.

Bortolotti, T., Boscari, S. and Danese, P. (2015) 'Successful lean implementation: Organizational culture and soft lean practices', International Journal of Production Economics, 160, pp. 182–201. doi: 10.1016/j.ijpe.2014.10.013.

de Souza, R. O., Ferenhof, H.A. and Forcellini, F.A. (2022) 'Systematic for process improvement using cyber-physical systems and Toyota Kata', in PLM 2021, 11–14 July 2021, pp. 447–460. Cham: Springer. doi: 10.1007/978-3-030-94335-6_32.

de Souza, R.O., Ribeiro, D.R.S., Ferenhof, H.A. and Forcellini, F.A. (2023b). Using Toyota Kata as a problem identification strategy. In Americas ICPR 2022. Cham: Springer. doi: 10.1007/978-3-031-36121-0_86

Eren, S. (2024) 'Toyota Kata as a scaffolding for human-centric manufacturing: Applying lean thinking for a digital and sustainable factory of the future', in FAIM 2023, Porto, 18–22 June 2023. Cham: Springer, pp. 380–387. doi: 10.1007/978-3-031-38165-2_45.

Faisal, A.M., Karthigeyan, L. and Suresh, G. (2024) 'Development of conceptual model for the introduction stage of lean intelligent manufacturing using simulation model based on the identification of the appropriate technologies of Industry 5.0', in ICIMES 2023, Hyderabad, 23-24 June 2023, vol. 372, pp. 237–246. Springer. doi: 10.1007/978-981-99-6774-2_22.

Fani, V., Bucci, I., Rossi, M. and Bandinelli, R. (2024) 'Lean and Industry 4.0 principles toward Industry 5.0: A conceptual framework and empirical insights from fashion industry', Journal of Manufacturing Technology Management, 35(9), pp. 122–141. doi: 10.1108/JMTM-11-2023-0509.

Frank, A.G., Thürer, M., Godinho Filho, M. and Marodin, G.A. (2024) 'Beyond Industry 4.0 – Integrating lean, digital technologies and people', International Journal of Operations and Production Management, 44(6), pp. 1109–1126. doi: 10.1108/IJOPM-01-2024-0069.

Houshmand, M. and Jamshidnezhad, B. (2006) 'An extended model of design process of lean production systems by means of process variables', Robotics and Computer-Integrated Manufacturing, 22(1), pp. 1–16. doi: 10.1016/j.rcim.2005.01.004.

Kaswan, M.S., Chaudhary, R., Garza-Reyes, J.A. and Singh, A. (2024) 'A review of Industry 5.0: From key facets to a conceptual implementation framework', International Journal of Quality & Reliability Management (ahead-of-print). https://doi.org/10.1108/IJQRM-01-2024-0030.

Köhler, C., Günes, M. and Schwinn, A. (2024) 'Human-centered optimization of assembly systems with the help of design thinking', Lecture Notes in Networks and Systems, 1059, pp. 249–257. doi: 10.1007/978-3-031-65411-4_30.

Kumar, V., Verma, P., Mittal, A., Gupta, P., Raj, R. and Kaswan, M.S. (2024a) 'Addressing the Kaizen business operations: The role of triple helix actors during COVID-19 outbreak', The TQM Journal, 36(6), pp. 1665–1694. doi: 10.1108/TQM-08-2023-0253.

Kumar, U., Kaswan, M.S., Kumar, R., Chaudhary, R., Garza-Reyes, J.A., Rathi, R. and Joshi, R. (2024b) 'A systematic review of Industry 5.0 from main aspects to the execution status', The TQM Journal, 36(6), pp. 1526–1549. doi: 10.1108/TQM-06-2023-0183.

Leng, J., Sha, W., Wang, B., Zheng, P., Zhuang, C., Liu, Q., Wuest, T., Mourtzis, D. and Wang, L. (2022) 'Industry 5.0: Prospect and retrospect', Journal of Manufacturing Systems, 65, pp. 279–295. doi: 10.1016/j.jmsy.2022.09.017.

Lu, Y. (2017) 'Industry 4.0: A survey on technologies, applications and open research issues', Journal of Industrial Information Integration, 6, pp. 1–10. doi: 10.1016/j.jii.2017.04.005.

Maddikunta, P.K.R., Pham, Q.-V., Prabadevi, B., Deepa, N., Dev, K., Gadekallu, T.R., Ruby, R. and Liyanage, M. (2022) 'Industry 5.0: A survey on enabling technologies and potential applications', Journal of Industrial Information Integration, 26, p. 100257. doi: 10.1016/j.jii.2021.100257.

A. Mittal, S. Sachan, V. Kumar, S. Vardhan, P. Verma, M.S. Kaswan, J.A. GarzaReyes, Essential organizational variables for the Implementation of Quality 4.0: Empirical evidence from the Indian furniture industry, The TQM Journal (2023), https://doi.org/10.1108/TQM-06-2023-0189. Ahead-of-Print

Moher, D., Liberati, A., Tetzlaff, J., Altman, D.G., Antes, G., Atkins, D., Barbour, V., Barrowman, N., Berlin, J.A., Clark, J., Clarke, M. and Cook, D. (2009). 'Preferred reporting items for systematic reviews and meta-analyses: The PRISMA statement', PLoS Medicine, 6(7), Article no. e1000097. doi: 10.1371/journal.pmed.1000097.

Moraes, A., Carvalho, A.M. and Sampaio, P. (2023) 'Lean and Industry 4.0: A review of the relationship, its limitations, and the path ahead with Industry 5', Machines, 11(4), Article 443. doi: 10.3390/machines11040443.

Ohno, T. (1988) Toyota Production System – Beyond Large-Scale Production. 1st edn. New York, NY: Productivity Press.

Rahardjo, B., Wang, F.-K., Lo, S.-C. and Chu, T.-H. (2024) 'A sustainable innovation framework based on Lean Six Sigma and Industry 5.0', Arabian Journal for Science and Engineering, 49(5), pp. 7625–7642. doi: 10.1007/s13369-023-08565-3.

Reverol, J. (2012) 'Creating an adaptable workforce: Using the Coaching Kata for enhanced environmental Performance', Environmental Quality Management, 22(2), pp. 19–31. doi: 10.1002/tqem.21324.

Ríos-Hernández, A., Mendoza-Gómez, J. and Valdez–de la Rosa, L.M. (2024) 'Emotional intelligence and analytical skill as human capital factors influencing competitiveness through analysis of Toyota Kata, Kaizen and Quality 4.0: Empirical study conducted during COVID-19', TQM Journal, 36(6), pp. 1402–1419. doi: 10.1108/TQM-05-2023-0129.

Rother, M. (2010) Toyota Kata: Managing People for Improvement, Adaptiveness and Superior Results. New York, NY: McGraw-Hill.

Rother, M. and Shook, J. (1998) Learning to See: Value Stream Mapping to Create Value and Eliminate Muda. Cambridge, MA: Lean Enterprise Institute.

Sabale, D.B., Kaswan, M.S. and Rathi, R. (2024) 'Development of circular economy model using Industry 4.0's machine learning techniques to improve performance metrics of MSMEs', International Journal on Interactive Design and Manufacturing, pp.1–19. doi: 10.1007/s12008-024-01782-6.

Sharma, S., Malik, A., Sharma, C., Batra, I., Kaswan, M.S. and Garza-Reyes, J.A. (2023). 'Adoption of industry 4.0 in different sectors: A structural review using natural language processing. International Journal on Interactive Design and Manufacturing, pp.1–23. doi:10.1007/s12008-023-01550-y.

Shehabuddeen, N., Probert, D. and Phaal, R. (2006) 'From theory to practice: Challenges in operationalising a technology selection framework', Technovation, 26(3), pp. 324–335.

Torre, N. and Bonamigo, A. (2024) 'Action research of lean 4.0 application to the maintenance of hydraulic systems in steel industry', Journal of Quality in Maintenance Engineering, 30(2), pp. 341–366. doi: 10.1108/JQME-06-2023-0058.

Tortorella, G.L. and Fettermann, D. (2018) 'Implementation of industry 4.0 and lean production in Brazilian manufacturing companies', International Journal of Production Research, 56(8), pp. 2975–2987. doi: 10.1080/00207543.2017.1391420.

Watson, R.T. and Webster, J. (2020) 'Analysing the past to prepare for the future: Writing a literature review a roadmap for release 2.0', Journal of Decision Systems, 29(3), pp. 129–147. doi: 10.1080/12460125.2020.1798591.

Womack, J.P. and Jones, D.T. (1996) Lean Thinking: Banish Waste and Create Wealth in Your Corporation. New York, NY: Simon & Schuster.

11 Industry 5.0 and Resource Complementarity

An Evaluation between Universities and Industries in Light of Value Co-Creation

Andrei Bonamigo, Steffan Macali Werner, José Eugênio Monteiro de Camargo Filho, and Jessica Borg

11.1 INTRODUCTION

Integrating university and industry can benefit both parties significantly (Ngoc et al., 2023). The positive impact is not limited solely to companies and universities but also extends to the community at large by bringing about positive environmental and social effects (Batra et al., 2024; Bocken et al., 2022), achieved through examples such as the sharing of knowledge and enhanced work readiness of graduates. The collaboration between university and industry benefits society by driving innovation, fostering economic development, educating qualified professionals, and addressing complex challenges (Borg & Scott-Young, 2020b; Bramwell et al., 2012). For organizations that engage in collaboration with universities, the benefits include innovation, access to qualified talents, solutions for complex challenges, access to global networks of academic research, and increased competitive advantage (Malodia et al., 2023; Kumar et al., 2023). Such partnerships are especially valuable in the context of Industry 5.0, which emphasizes technologies that not only drive innovation but also enhance sustainability, resilience, and human-centricity (Hussain et al., 2023; Javaid et al., 2022; Nayeri et al., 2023). In addition, Industry 5.0 promotes and facilitates the use of new technologies, fostering engagement among stakeholders and the exchange of information (Bonamigo et al., 2024).

This study aims to explore the relationship between universities and the industries from a value co-creation perspective based on Industry 5.0 technologies. Through this examination, the focus of our study is not only to gain insights but also to identify potential strategies for fostering greater collaboration, thereby contributing to social development and creating tangible value for the industry. Through this study,

DOI: 10.1201/9781003535188-11

we sought to understand the current stage of interaction between university and industry. As such, our key research question was: *What is the level of interaction currently being undertaken between universities and industries?*

11.2 METHOD

This chapter endeavored to synthesize academic insights into the intricate dynamics between universities and industries, striving to cultivate a comprehension of this relationship. The initial step involved a systematic literature review (SLR) to identify the current state of knowledge on the subject, drawing on the framework provided by Ferenhof and Fernandes (2016).

Subsequently, a content analysis, following the stages of pre-analysis, exploration of material or coding, and treatment of results, inference, and interpretation, as outlined by Bardin (2011), was carried out, as described below.

11.2.1 Systematic Literature Review – SLR

To accomplish this objective, we adhered to the systematic search flow (SSF) method outlined by Ferenhof and Fernandes (2016) for conducting the SLR. The SSF method comprises four phases and eight activities.

Following this, we provide concise descriptions of the respective phases and activities (Ferenhof & Fernandes, 2016). The Research Protocol Definition phase centers on delineating the study's object and is divided into five sub-stages. The search strategy involves a series of procedures that outline the selection of search engines, the definition of inclusion and exclusion criteria, and the retrieval of online information.

First, we formulated a research plan encompassing the questions we sought to address, which were as follows:

a. What is the current nature of the relationship between universities and industries?
b. What are the inhibitors, obstacles, barriers, or facilitators for effective co-creation of value between these agents?
c. What possible solutions are being employed to enhance this interaction?

This also encompassed the selection of keywords and a set of inclusion and exclusion criteria. Subsequently, we opted for the research query *((university) AND ("innovation ecosystem" OR "value co-creation" OR cooperation OR collaboration) AND ("industry 5.0" OR "human-centred" OR "human-centered" OR "socio-technical") AND (technological OR technology OR "digital transformation"))*. The inclusion criteria comprised empirical papers that were peer-reviewed, written in English, Spanish, or Portuguese, and indexed in the databases: Compendex, Ebesco, Scopus, Emerald, Science Direct, ProQuest, and Web of Knowledge.

In line with the SSF method's recommendation, we generated a "knowledge matrix" in a data sheet containing criteria to recognize the relationship between universities and the industries from the value co-creation perspective. Subsequently,

TABLE 11.1
Resulting Bibliographic Portfolio

Database	Number of Documents Found
ISI Web of Science	818
Compendex	800
Science Direct	400
Emerald	351
Ebesco	249
Scopus	73
ProQuest	17
Subtotal	3108
Duplicates	(−924)
Total	**2184**

a member of the research team accessed the databases and conducted a search using the predefined combination of keywords.

We examined this combination within the title, keywords, and abstract. The database search was conducted on 21 August 2023, returning 3108 documents. Following that, we managed the documents by importing references into reference management software, subsequently eliminating duplicates, resulting in 2184 documents (Table 11.1).

Scholars have begun to appraise the use of ChatGPT and AI in academic research, including the use of ChatGPT when conducting literature reviews (Alshami et al., 2023; Haman & Školník, 2023). Haman and Školník (2023) found that ChatGPT is not suitable for conducting a literature search due to its tendency to generate non-existent papers. Scholars have found that ChatGPT has limitations in extracting articles (Alshami et al., 2023).

Hence, we did not use the AI program to search but rather to ascertain whether the academic papers generated from a search strategy were indeed relevant to our topic and area of interest (i.e., whether they explored/elicited a correlation between industries and universities). This is in alignment with the findings of other scholars (Alshami et al., 2023), who, in investigating ChatGPT's capabilities in assisting with systematic reviews of academic literature, found that "ChatGPT exhibits exceptional performance in filtering and categorizing relevant articles, leading to significant time and effort savings" (p. 1).

In the fifth step, we compiled the bibliographic portfolio for analysis by exporting author (year), title, and journal information from the reference management software to a spreadsheet. As the sixth step, we meticulously reviewed each data entry, editing the content and coding it according to the criteria outlined in Table 11.2.

Seventh, after the synthesis stage, the data from each author were consolidated into a single spreadsheet. This review allowed us to categorize the findings into themes, thereby aiding in elucidating insights into drivers, key factors, inhibitors, obstacles, barriers, or facilitators for the effective co-creation of value between these entities.

TABLE 11.2
Resulting Bibliographic Portfolio

Code	Author (Year)	Title
A1	Abbate et al. (2019)	Knowledge Co-Creation in Open Innovation Digital Platforms: Processes, Tools and Services
A2	Abdullah et al. (2020)	Industry Revolution 4.0: The Readiness of Graduates of Higher Education Institutions for Fulfilling Job Demands
A3	Abrantes et al. (2021)	An Augmented Infocommunication Model for Unified Communications in Situational Contexts of Collaboration
A4	Adeyeye et al. (2021)	Socio-Technical Viability Framework for Micro Hydropower in Group Water-Energy Schemes
A5	Agapie et al. (2022)	Using HCI in Cross-Disciplinary Teams: A Case Study of Academic Collaboration in HCI-Health Teams in the US Using a Team Science Perspective
A6	Ajibade et al. (2023)	Machine Learning Applications in Renewable Energy (MLARE) Research: A Publication Trend and Bibliometric Analysis Study (2012–2021)
A7	Albers et al. (2018)	Connected Creativity – A Human Centered Community Innovation Platform in the Context of Product Generation Engineering
A8	Albino et al. (2014)	Understanding the Development Trends of Low-Carbon Energy Technologies: A Patent Analysis
A9	Alreshidi et al. (2017)	Factors for Effective BIM Governance
A10	Al-Tabbaa and Ankrah (2019)	"Engineered" University-Industry Collaboration: A Social Capital Perspective
A11	Amanatidou et al. (2016)	Strategies for Emerging Research and Innovation Futures
A12	Andersen et al. (2011)	Designing for Collaborative Interpretation in Telemonitoring: Re-Introducing Patients as Diagnostic Agents
A13	Ardito et al. (2019)	The Role of Universities in the Knowledge Management of Smart City Projects
A14	Bakir and Dahlan (2023)	Higher Education Leadership and Curricular Design in Industry 5.0 Environment: A Cursory Glance
A15	Bjerregaard (2010)	Industry and Academia in Convergence: Micro-Institutional Dimensions of R&D Collaboration
A16	Charungkaittikul et al. (2021)	Fostering Self-Direction in Foresight and Skills Anticipation of Undergraduate Students Through University-Industry Engagement: Future Signs for Thailand Project
A17	DeWaters and Kotla (2023)	Using an Open-Ended Socio-Technical Design Challenge for Entrepreneurship Education in a First-Year Engineering Course
A18	Doganova (2013)	Transfer and Exploration: Two Models of Science-Industry Intermediation
A19	Eriksson et al. (2022)	Experiences in Running a Professional Course on Digitally-Enabled Production in Collaboration between Three Swedish Universities
A20	Fernandes and O'Sullivan (2021)	Benefits Management in University-Industry Collaboration Programs

(Continued)

TABLE 11.2 *(Continued)*
Resulting Bibliographic Portfolio

Code	Author (Year)	Title
A21	Ferraris et al. (2020)	The Role of Universities in the Smart City Innovation: Multistakeholder Integration and Engagement Perspectives
A22	Herro et al. (2021)	Exploring Graduate Students' Collaborative Problem-Solving in Engineering Design Tasks
A23	Högfeldt et al. (2018)	Understanding Engineering Education Change with the Introduction of Challenge Driven Education in Tanzania
A24	Hussain et al. (2023)	Next Generation Employability and Career Sustainability in the Hospitality Industry 5.0
A25	Ihsen et al. (1998)	Creativity, Complexity and Self-Similarity: The Vision of the Fractal University
A26	Joshi and White (2020)	Design Thinking Approach to Identify Barriers to Engineering Education Reform in India
A27	Kadir et al. (2019)	Current Research and Future Perspectives on Human Factors and Ergonomics in Industry 4.0
A28	Kruesi et al (2020)	A Knowledge Management System Framework for an Open Biomedical Repository: Communities, Collaboration and Corroboration
A29	Kuys et al. (2021)	A Case Study of University–Industry Collaboration for Sustainable Furniture Design
A30	Liwång (2022)	Defense Development: The Role of Co-Creation in Filling the Gap between Policy-Makers and Technology Development
A31	Martínez-Cerdá et al. (2018)	Opening the Black-Box in Lifelong E-Learning for Employability: A Framework for a Socio-Technical E-Learning Employability System of Measurement (STELEM)
A32	Martínez-Cerdá et al. (2018)	Promoting Collaborative Skills in Online University: Comparing Effects of Games, Mixed Reality, Social Media, and Other Tools for ICT-Supported Pedagogical Practices
A33	Muxfeldt et al. (2015)	Developing New Application Fields for Industrial Robots – Four Examples for Academia-Industry Collaboration
A34	Noor and Venneri (2008)	ISE: Intelligent Synthesis Environment for Future Aerospace Systems
A35	Pittaway et al. (2020)	University Spaces for Entrepreneurship: A Process Model

Subsequently, all authors collaborated and collectively identified themes for each entry. These discussions also resulted in further reductions in the number of papers. In the end, 35 papers were included in the review. The eighth stage, the final phase of our review process, focused on writing up the findings.

The content analysis comprised (1) pre-analysis; (2) exploration of material or coding; and (3) treatment of results, inference, and interpretation, following the guidelines outlined by Bardin (2011), as detailed below. For this stage, while reviewing the portfolio of papers, the authors examined the references in those 35 documents and discovered an additional 12 works that were aligned with the theme and were subsequently included in the final bibliographic portfolio. Subsequently, the 47 documents were coded and analyzed according to the content analysis criteria specified by Bardin (2011).

Following that, individual data were synthesized into a single spreadsheet. Subsequently, each researcher independently reviewed the consolidated sheet to ensure consistency regarding the coding of the context unit and record unit. Our varied interpretations were shared and discussed during our discussion cycles. These cycles of discussion led to a further reduction in the number of papers. In the end, 35 papers formed the foundation for analysis (Table 11.2).

Then, the final stage of our review process was devoted to the write-up of the findings. The united analysis was defined as "*a posteriori*".

11.3 FINDINGS

Through content analysis, we organized the findings as elicited from the review of articles into three distinct topics/themes – *knowledge management*, *value co-creation*, and *social responsibility*. This is shown in Table 11.3. This decision was made to direct the readers' attention to three specific subjects that have a direct correlation with the discussed topic.

TABLE 11.3
Resulting Bibliographic Portfolio

Record Unit	Context Unit	Freq.
Knowledge Management	A2 – Integrating learning factory principles into university curricula serves as a means of better-preparing students for the job market. A3 – Advancements in communication are bringing universities and industries closer together, fostering the generation of new technologies. A5 – Collaborative endeavors between industry and universities in research funding. A10 – How to effectively share knowledge and technology between entities? A12 – Interdisciplinary curriculum designed to apply a diverse range of distinct knowledge and enhance students' preparation for the job market. A13 – Universities have a central role in generating, preserving, and disseminating knowledge. A18 – Businesses capitalizing on knowledge and technologies originating from universities. A19 – Attraction of qualified professionals for a career in teaching. A21 – Universities as generators of knowledge for practical application and sharing with industry. A24 – The changes in education in response to industry transformation. A32 – The current relationship of Information and Communication Technologies with students and industries. A35 – University reform to attract entrepreneurial students.	12

(Continued)

TABLE 11.3 *(Continued)*
Resulting Bibliographic Portfolio

Record Unit	Context Unit	Freq.
Value Co-creation	A1 – Initiatives to facilitate connectivity among universities, research centers, and companies. A4 – The collaboration between industry and universities, with a social perspective, brings about positive impacts. A9 – Establishing open communication channels between industry and universities to seek effective solutions to challenges. A14 – Higher Education Institutions (HEIs) are intensifying their engagement with industry. A15 – Cultural differences presenting challenges in the relationships between academia and industries. A16 – Collaboration through the generation of market analysis. A20 – Convergence and clear definitions of roles as a means to achieving positive outcomes. A23 – Value co-creation with different cultures and realities for diverse perspectives. A28 – Collaboration reduces duplication of efforts and removes barriers in a more effective way. A29 – Collaboration with a focus on the client to improve customer services. A30 – Collaboration ensures that policies and social values are taken into account in the development of new technologies. A33 – Challenges in the exchange of information between university and industry. A34 – Initiatives to enhance value co-creation.	13
Social Responsibility	A6 – Reconnaissance by governments of the potential progress stemming from the industry-university relationship. A7 – Researchers need to understand the potential they have to find solutions and approach industries to assist them. A8 – Governments act as catalysts for partnerships between universities and industries due to the significance of sustainable development. A11 – Research funding and university accountability. A17 – Curriculum with socio-technical design challenges for students to develop skills for the job market. A22 – Educational institutions focusing more on practical problem-solving solutions to achieve greater impact in society. A25 – The need for the adaptation of universities to the new automation and information networks in the industry. A26 – Low qualification after graduation due to a lack of proper dissemination of opportunities offered in universities. A27 – The importance of universities in enhancing the ergonomics of workers, particularly at lower levels. A31 – Digital education that empowers students.	10

11.3.1 Knowledge Management

Knowledge management (KM) is a strategic approach for acquiring, sharing, and leveraging both individual and collective knowledge (De Bem Machado et al., 2022). Effective KM, driven by innovative strategies, adaptability, and technologies like neural networks, is crucial for enhancing competitiveness, fostering innovation, and ensuring the continuous development of companies (Ferenhof et al., 2024). In this context, Industry 5.0 technologies can accelerate and streamline knowledge-sharing by integrating advanced, human-centered tools, promoting a more agile and collaborative information flow between people and systems. KM platforms also enable real-time access to information, updates, and decision-making (Bonamigo et al., 2024; Hossain et al., 2022).

Sun et al. (2022) stress that for KM to be successfully implemented, its importance in optimizing organizational performance must be recognized, with the use of structural equation modeling for in-depth analysis. Integrating KM into the supply chain is also key. However, challenges such as misunderstanding the KM concept, inefficient knowledge transfer, and conflicts due to demographic and cultural differences may arise (Pepple et al., 2022). Kaswan et al. (2024) highlight the need to learn from human-robot interactions to improve services and enhance organizational efficiency and sustainability, allowing for faster responsiveness in various situations. This demonstrates the innovation potential in industrial and service sectors through university collaboration.

11.3.2 Value Co-Creation

Value co-creation is a collaborative process in which various stakeholders work together to optimize resources, promote efficiency, and generate value throughout the production chain (Bonamigo et al., 2023). From this perspective, Kumar et al. (2024) present Industry 5.0 as a collaborative approach based on the synergy between humans and robots to drive innovations and overcome practical challenges faced by organizations. Value co-creation with universities and industries can foster this collaboration. Strategic partnerships and active collaborations facilitate and motivate value co-creation between industry and academic institutions (Pocol et al., 2022). Industry 5.0 contributes to the operationalization of co-creation by promoting sustainable and resilient system development and utilizing contextual information for decision-making (Bakir & Dahlan, 2023).

Value co-creation enables different teams to collaborate in driving innovation by sharing big data among multiple players, exploring it to provide new services, solutions, or knowledge to address crises (Bonamigo et al., 2023; Sharma et al., 2023). It is crucial to consider the needs of different stakeholders in the collaboration process to ensure win-win solutions for both academia and industry. For instance, while universities often aim to disseminate knowledge, industries prioritize business confidentiality (Kumar et al., 2024; Muxfeldt et al., 2015).

Partnerships between industries and universities, when viewed from a social perspective, can generate significant positive impacts in various areas, such as advancing technological development and increasing engagement of both employees and students (Adeyeye et al., 2021; Kumar et al., 2024). However, universities must adapt to changes in the professional landscape shaped by automation and information networks in the industry (Ihsen et al., 1998; Sabale et al., 2024). Both parties must make efforts to ensure the relationship is collaborative and productive.

Albers et al. (2018) highlight the importance of researchers developing insights and empathy for the needs and challenges faced by industry partners. Open communication channels between industries and universities are crucial for enabling researchers to understand the sector's challenges and seek effective solutions (Alreshidi et al., 2017; Borg & Scott-Young, 2020a). Furthermore, universities play a key role in improving worker ergonomics, particularly at lower levels, while also addressing human challenges (Kadir et al., 2019; Sabale et al., 2024).

11.3.3 Social Responsibility

Social responsibility refers to the commitment of organizations to adopt ethical practices and contribute to societal well-being (Upadhyay et al., 2021). In this context, digital transformation and Industry 5.0 enable universities to prioritize social responsibility by actively engaging with local communities, sharing knowledge, continuously monitoring and managing impact, and obtaining a "social license to operate" through effective engagement and impact management strategies (Chen & Vanclay, 2021).

Developing critical thinking skills in students and professionals through methods such as task analysis and problem-solving plays a key role in promoting social responsibility. This empowers individuals to find solutions, demonstrate their abilities, and contribute to societal development (Renatovna & Renatovna, 2021; Torre & Bonamigo, 2024). However, from an Industry 5.0 perspective, implementing these methods faces challenges, including the need for family involvement from early childhood, overcoming deeply rooted traditions, and creating a supportive educational environment. This highlights the complexity of fostering critical thinking in students and emphasizes the shared responsibility between families and educators (Renatovna & Renatovna, 2021). Additionally, ethical risks related to artificial technologies require ongoing monitoring and evaluation within the scope of social responsibility (Kumar et al., 2024).

11.3.4 Conclusion, Limitations, and Future Research Agenda

This study aimed to explore the relationship between universities and the industries from a value co-creation perspective based on Industry 5.0 technologies. The findings of this research emphasize the multi-faceted relationship between universities and industries, highlighting both the positive impacts and challenges associated with their collaboration for resource complementarity. The technologies of Industry 5.0 support the integration between different parts, facilitating the sharing of information and knowledge. New skills are required for the operationalization of Industry 5.0. Although it facilitates orchestration and mediates the complementarity of resources, its implementation demands specific knowledge that can be developed through collaboration between universities and industries.

REFERENCES

Abbate, T., Codini, A. P., & Aquilani, B. (2019). Knowledge Co-Creation in Open Innovation Digital Platforms: Processes, Tools and Services. *Journal of Business & Industrial Marketing, 34*(7), 1434–1447.

Abdullah, Q. A., Humaidi, N., & Shahrom, M. (2020). Industry Revolution 4.0: The Readiness of Graduates of Higher Education Institutions for Fulfilling Job Demands. *Romanian Journal of Information Technology & Automatic Control/Revista Română de Informatică şi Automatică, 30*(2), 15–26.

Abrantes, C., Mealha, Ó, Gomes, D., Paulo Barraca, J., & Viana-Ferreira, C. (2021). An Augmented Infocommunication Model for Unified Communications in Situational Contexts of Collaboration. *Interdisciplinary Journal of Information, Knowledge & Management, 16*, 77–99.

Adeyeye, K., Gallagher, J., McNabola, A., Ramos, H. M., & Coughlan, P. (2021). Socio-Technical Viability Framework for Micro Hydropower in Group Water-Energy Schemes. *Energies, 14*(14), 4222.

Agapie, E., Haldar, S. & Poblete, S.G., (2022). Using HCI in Cross-Disciplinary Teams: A Case Study of Academic Collaboration in HCI-Health Teams in the US Using a Team Science Perspective. *Proceedings of the ACM on Human-Computer Interaction*, 6(CSCW2), pp.1–35.

Ajibade, S. S. M., Bekun, F. V., Adedoyin, F. F., Gyamfi, B. A., & Adediran, A. O. (2023). Machine Learning Applications in Renewable Energy (MLARE) Research: A Publication Trend and Bibliometric Analysis Study (2012–2021). *Clean Technologies, 5*(2), 497–517.

Albers, A., Maul, L., Heismann, R., & Bursac, N. (2018). Connected Creativity – A Human Centered Community Innovation Platform in the Context of Product Generation Engineering. *Design Science, 4*, e6.

Albino, V., Ardito, L., Dangelico, R. M., & Petruzzelli, A. M. (2014). Understanding the Development Trends of Low-Carbon Energy Technologies: A Patent Analysis. *Applied Energy, 135*, 836–854.

Alreshidi, E., Mourshed, M., & Rezgui, Y. (2017). Factors for Effective BIM Governance. *Journal of Building Engineering, 10*, 89–101.

Alshami, A., Elsayed, M., Ali, E., Eltoukhy, A. E. E., & Zayed, T. (2023). Harnessing the Power of ChatGPT for Automating Systematic Review Process: Methodology, Case Study, Limitations, and Future Directions. *Systems, 11*, 351.

Al-Tabbaa, O. & Ankrah, S., (2019). Engineered' University-Industry Collaboration: A Social Capital Perspective. *European Management Review, 16*(3), 543–565.

Amanatidou, E., Saritas, O., & Loveridge, D. (2016). Strategies for Emerging Research and Innovation Futures. *Foresight, 18*(3), 253–275.

Andersen, T., Bjørn, P., Kensing, F., & Moll, J. (2011). Designing for Collaborative Interpretation in Telemonitoring: Re-Introducing Patients as Diagnostic Agents. *International Journal of Medical Informatics, 80*(8), e112–e126.

Ardito, L., Ferraris, A., Petruzzelli, A. M., Bresciani, S., & Del Giudice, M. (2019). The Role of Universities in the Knowledge Management of Smart City Projects. *Technological Forecasting and Social Change, 142*, 312–321.

Bakir, A., & Dahlan, M. (2023). Higher Education Leadership and Curricular Design in Industry 5.0 Environment: A Cursory Glance. *Development and Learning in Organizations: An International Journal, 37*(3), 15–17.

Bardin, L. (2011). Análise de conteúdo. Edições 70. *Lisboa. Portugal.*

Batra, I., Sharma, C., Malik, A., Sharma, S., Kaswan, M. S., & Garza-Reyes, J. A. (2024). Industrial Revolution and Smart Farming: A Critical Analysis of Research Components in Industry 4.0. *The TQM Journal*, (ahead-of-print).

Bjerregaard, T. (2010). Industry and Academia in Convergence: Micro-Institutional Dimensions of R&D Collaboration. *Technovation, 30*(2), 100–108.

Bocken, N. M., Niessen, L., & Short, S. W. (2022). The Sufficiency-Based Circular Economy—An Analysis of 150 Companies. *Frontiers in Sustainability*, *3*, 899289.

Bonamigo, A., de Azeredo, R. R., Monteiro de Camargo Filho, J. E., & de Souza Andrade, H. (2023). Lean 4.0 in the Value Co-Creation in Agro-Industrial Services: An Agenda for Future Studies for the Efficient Resource Use. *Systems Research and Behavioral Science*, *1*, 1–16.

Bonamigo, A., Werner, S. M., & de Oliveira, M. S. (2024). Toward Lean Service 5.0: Evaluation of Future Perspectives and Trends. *Revista Meta: Avaliação*, *16*(50), 180–210.

Borg, J., & Scott-Young, C. M. (2020a). Priming the Project Talent Pipeline: Examining Work Readiness in Undergraduate Project Management Degree Programs. *Project Management Journal*, *51*(2), 165–180.

Borg, J., & Scott-Young, C. M. (2020b). Employers' Perspectives on Work Readiness in Construction: Are Project Management Graduates Hitting the Ground Running? *International Journal of Managing Projects in Business*, *13*(6), 1363–1379.

Bramwell, A., Hepburn, N., & Wolfe, D. A. (2012). Growing innovation ecosystems: University-industry knowledge transfer and regional economic development in Canada. Final *Report to the Social Sciences and Humanities Research Council of Canada*, 62.

Charungkaittikul, S., Areesophonpichet, S., & Malaiwong, W. (2021). Fostering Self-Direction in Foresight and Skills Anticipation of Undergraduate Students Through University-Industry Engagement: Future Signs for Thailand Project. *International Journal of Adult Education and Technology (IJAET)*, *12*(3), 47–61.

Chen, C., & Vanclay, F. (2021). Transnational Universities, Host Communities and Local Residents: Social Impacts, University Social Responsibility and Campus Sustainability. *International Journal of Sustainability in Higher Education*, *22*(8), 88–107.

De Bem Machado, A., Secinaro, S., Calandra, D., & Lanzalonga, F. (2022). Knowledge Management and Digital Transformation for Industry 4.0: A Structured Literature Review. *Knowledge Management Research & Practice*, *20*(2), 320–338.

DeWaters, J., & Kotla, B. (2023, May). Using an Open-Ended Socio-Technical Design Challenge for Entrepreneurship Education in a First-Year Engineering Course. *Frontiers in Education*, *8*, 1198161.

Doganova, L. (2013). Transfer and Exploration: Two Models of Science-Industry Intermediation. *Science and Public Policy*, *40*(4), 442–452.

Eriksson, K. M., Chirumalla, K., Myrelid, P., Ericsson, M., Granlund, A., Håkansson, L., & Johansson, D. (2022). Experiences in Running a Professional Course on Digitally-Enabled Production in Collaboration between Three Swedish Universities. *Advances in Transdisciplinary Engineering*, *21*, 653–664.

Ferenhof, H. A., Bonamigo, A., Rosa, L. G., & Vieira, T. C. (2024). Theoretical B2B Knowledge Management Framework Focused on Value Co-Creation. *VINE Journal of Information and Knowledge Management Systems*, *54*(2), 424–451.

Ferenhof, H. A., & Fernandes, R. F. (2016). Demystifying the Literature Review as Basis for Scientific Writing: SSF Method. *Revista ACB*, *21*(3), 550–563.

Fernandes, G., & O'Sullivan, D. (2021). Benefits Management in University-Industry Collaboration Programs. *International Journal of Project Management*, *39*(1), 71–84.

Ferraris, A., Belyaeva, Z., & Bresciani, S. (2020). The Role of Universities in the Smart City Innovation: Multistakeholder Integration and Engagement Perspectives. *Journal of Business Research*, *119*, 163–171.

Haman, M., & Školník, M. (2023). Using ChatGPT to Conduct a Literature Review. *Accountability in Research*, 1–3.

Herro, D., McNeese, N., O'Hara, R., Frady, K., & Switzer, D. (2021). Exploring Graduate Students' Collaborative Problem-Solving in Engineering Design Tasks. *Journal of Engineering Design*, *32*(9), 496–516.

Högfeldt, A. K., Gumaelius, L., Lantz, A., & Lujara, S. (2018, April). Understanding Engineering Education Change with the Introduction of Challenge Driven Education in Tanzania. In *2018 IEEE Global Engineering Education Conference (EDUCON)*, 1335–1343.

Hossain, M. B., Nassar, S., Rahman, M. U., Dunay, A., & Illés, C. B. (2022). Exploring the Mediating Role of Knowledge Management Practices to Corporate Sustainability. *Journal of Cleaner Production*, *374*, 133869.

Hussain, S., Singh, A. M., Mohanty, P., & Gavinolla, M. R. (2023). Next Generation Employability and Career Sustainability in the Hospitality Industry 5.0. *Worldwide Hospitality and Tourism Themes*, *15*(3), 308–321.

Ihsen, S., Isenhardt, I., & Sanchez, U. S. D. (1998). Creativity, Complexity and Self-Similarity: The Vision of the Fractal University. *European Journal of Engineering Education*, *23*(1), 13–22.

Javaid, M., Haleem, A., Singh, R. P., Suman, R., & Gonzalez, E. S. (2022). Understanding the Adoption of Industry 4.0 Technologies in Improving Environmental Sustainability. *Sustainable Operations and Computers*, *3*, 203–217.

Joshi, R., & White, J. R. (2020, June). Design Thinking Approach to Identify Barriers to Engineering Education Reform in India. In *2020 ASEE Virtual Annual Conference Content Access*.

Kadir, B. A., Broberg, O., & da Conceicao, C. S. (2019). Current Research and Future Perspectives on Human Factors and Ergonomics in Industry 4.0. *Computers & Industrial Engineering*, *137*, 106004.

Kaswan, M. S., Chaudhary, R., Garza-Reyes, J. A. and Singh, A. (2024), "A review of Industry 5.0: from key facets to a conceptual implementation framework", *International Journal of Quality & Reliability Management*, Vol. ahead-of-print No. ahead-of-print. https://doi.org/10.1108/IJQRM-01-2024-0030

Kruesi, L., Burstein, F., & Tanner, K. (2020). A Knowledge Management System Framework for an Open Biomedical Repository: Communities, Collaboration and Corroboration. *Journal of Knowledge Management*, *24*(10), 2553–2572.

Kumar, U., Kaswan, M. S., Kumar, R., Chaudhary, R., Garza-Reyes, J. A., Rathi, R., & Joshi, R. (2023). A Systematic Review of Industry 5.0 from Main Aspects to the Execution Status. *The TQM Journal*, *36*(6), 1526–1549.

Kumar, V., Verma, P., Mittal, A., Gupta, P., Raj, R., & Kaswan, M. S. (2024). Addressing the Kaizen Business Operations: The Role of Triple Helix actors during COVID-19 Outbreak. *The TQM Journal*, *36*(6), 1665–1694.

Kuys, J., Al Mahmud, A., & Kuys, B. (2021). A Case Study of University–Industry Collaboration for Sustainable Furniture Design. *Sustainability*, *13*(19), 10915.

Liwång, H. (2022). Defense Development: The Role of Co-Creation in Filling the Gap between Policy-Makers and Technology Development. *Technology in Society*, *68*, 101913.

Malodia, S., Chauhan, C., Jabeen, F. and Dhir, A. (2023), "Antecedents and consequences of open innovation: a conceptual framework", *International Journal of Entrepreneurial Behavior & Research*, Vol. ahead-of-print No. ahead-of-print. https://doi.org/10.1108/IJEBR-03-2022-0235

Martínez-Cerdá, J.F., Torrent-Sellens, J., González-González, I. and Ficapal-Cusí, P., (2018b). Opening the Black-Box in Lifelong E-Learning for Employability: A Framework for a Socio-Technical E-learning Employability System of Measurement (STELEM). *Sustainability*, *10*(4), 1014.

Muxfeldt, A., Kubus, D., & Wahl, F. M. (2015, September). Developing New Application Fields for Industrial Robots – Four Examples for Academia-Industry Collaboration. In *2015 IEEE 20th Conference on Emerging Technologies & Factory Automation* (ETFA), 1–7.

Nayeri, S., Sazvar, Z., & Heydari, J. (2023). Towards a Responsive Supply Chain Based on the Industry 5.0 Dimensions: A Novel Decision-Making Method. *Expert Systems with Applications*, *213*, 119267.

Ngoc, N. M., Tien, N. H., Hieu, V. M., & Trang, T. T. T. (2023). Sustainable integration in Vietnam's tourism industry. *World review of entrepreneurship management and sustainable development*, 3, 1-16.

Noor, A. K., & Venneri, S. L. (2008). ISE: Intelligent Synthesis Environment for Future Aerospace Systems. *IEEE Aerospace and Electronic Systems Magazine*, *23*(4), 31–44.

Pepple, D., Makama, C., & Okeke, J. P. (2022). Knowledge Management Practices: A Public Sector Perspective. *Journal of Business Research*, *153*, 509–516.

Pittaway, L., Aissaoui, R., Ferrier, M., & Mass, P. (2020). University Spaces for Entrepreneurship: A Process Model. *International Journal of Entrepreneurial Behavior & Research*, *26*(5), 911–936.

Pocol, C. B., Stanca, L., Dabija, D. C., Pop, I. D., & Mişcoiu, S. (2022). Knowledge Co-Creation and Sustainable Education in the Labor Market-Driven University–Business Environment. *Frontiers in Environmental Science*, *10*, 781075.

Renatovna, A. G., & Renatovna, A. S. (2021). Pedagogical and Psychological Conditions of Preparing Students for Social Relations on the Basis of the Development of Critical Thinking. *Psychology and Education Journal*, *58*(2), 4889–4902.

Sabale, D. B., Kaswan, M. S., & Rathi, R. (2024). Development of Circular Economy Model Using Industry 4.0's Machine Learning Techniques to Improve Performance Metrics of MSMEs. *International Journal on Interactive Design and Manufacturing (IJIDeM)*, 1–19.

Sharma, S., Malik, A., Sharma, C., Batra, I., Kaswan, M. S., & Garza-Reyes, J. A. (2023). Adoption of Industry 4.0 in Different Sectors: A Structural Review Using Natural Language Processing. *International Journal on Interactive Design and Manufacturing (IJIDeM)*, 1–23.

Sun, Y., Shahzad, M., & Razzaq, A. (2022). Sustainable Organizational Performance through Blockchain Technology Adoption and Knowledge Management in China. *Journal of Innovation & Knowledge*, *7*(4), 100247.

Torre, N. M. D. M., & Bonamigo, A., (2024). Action Research of Lean 4.0 Application to the Maintenance of Hydraulic Systems in Steel Industry. *Journal of Quality in Maintenance Engineering*, *30*(2), 341–366.

Upadhyay, A., Mukhuty, S., Kumar, V., & Kazancoglu, Y., (2021). Blockchain Technology and the Circular Economy: Implications for Sustainability and Social Responsibility. *Journal of Cleaner Production*, 293, 126130.

12 Harnessing Industry 5.0 for Sustainable Urban Logistics

Integrating Advanced Technologies with Green Practices

Ish Kumar and Chidambara

12.1 INTRODUCTION

Industry 5.0 represents the subsequent stage in the evolution of industrial practices, building on the advancements made through Industry 4.0. Industry 4.0 emphasizes automation, connectivity, in addition to the integration of cyber-physical systems, IoT, and big data in creating intelligent factories, whereas the Industry 5.0 model introduces a more human-centric framework. Industry 5.0 integrates human innovation and creativity alongside advanced technologies like AI, IoT, and robotics to enable highly customized, effective, and sustainable production processes (Nahavandi, 2019). Industry 4.0's emphasis on automation and digital interconnectivity transforms manufacturing by introducing intelligent systems capable of predictive maintenance, real-time decision-making, and optimized production (Tao et al., 2018; Zhong et al., 2017). However, this shift raises concerns about the displacement of human workers and the environmental impacts of mass automation. Industry 5.0 addresses these concerns by emphasizing the role of humans in decision-making and creativity, fostering collaborative interactions between humans and machines (Nahavandi, 2019). This shift marks a transition from automation-driven processes to a model where human ingenuity and technology converge to create resilient and adaptable industrial systems (Ivanov et al., 2017). Recently, Industry 5.0 has garnered attention for its ability to tackle sustainability-related challenges. By integrating AI, IoT, and robotics with human oversight, Industry 5.0 offers the opportunity to optimize resource use, cut down on waste, and reduce the environmental impact of industrial operations. This is especially relevant in sectors like urban logistics, where sustainability concerns are becoming increasingly critical (Dablanc & Rodrigue, 2014). Urban logistics, especially in e-commerce logistics and last-mile delivery operations, faces significant challenges, including increased greenhouse gas emissions, traffic congestion, and inefficiencies in delivery operations (Holguín-Veras et al., 2016). By

 DOI: 10.1201/9781003535188-12

applying Industry 5.0 technologies, urban logistics can be optimized to reduce its environmental impact, improve operational efficiency, and contribute to the growth of sustainable urban settings (DHL, 2017). The role of Industry 5.0 in urban logistics is critical as it facilitates the incorporation of green logistics practices, including the adoption of electric vehicles (EVs), AI-driven route optimization, and the creation of urban consolidation centers (UCCs) to alleviate congestion (Morganti & Browne, 2018). These technologies, when combined with human ingenuity, allow for a flexible and adaptive logistics system that is both efficient and environmentally conscious (Saberi et al., 2019).

12.2 LITERATURE REVIEW

Industry 5.0 has emerged as a necessary progression from Industry 4.0, which prioritized automation and leveraged technologies like IoT, AI, and big data to improve production efficiency (Kaswan et al., 2024). Though Industry 4.0 brought significant changes to manufacturing, its focus on automation led to concerns over job loss and environmental sustainability. These concerns have facilitated the growth of Industry 5.0, which emphasizes human-machine collaboration and sustainable practices (Nahavandi, 2019). In Industry 5.0, human creativity and intelligence play a key role in production processes, with technologies like AI and IoT complementing rather than replacing human skills. This human-centric approach has significant implications for urban logistics, particularly in the context of sustainability (Viu-Roig & Alvarez-Palau, 2020). Urban logistics, defined as the process of moving goods within urban environments, faces numerous challenges, including traffic congestion, air pollution, and inefficiencies in last-mile delivery (Rodrigue et al., 2016). One of the main areas where Industry 5.0 can contribute to more sustainable urban logistics is through the integration of green technologies. EVs, for example, can reduce the carbon footprint of logistics operations, while AI-driven route optimization can help minimize fuel consumption and emissions (Morganti & Browne, 2018). Additionally, UCCs equipped with IoT technology have demonstrated the ability to cut down the number of delivery vehicles on the road, which helps reduce traffic congestion and pollution (Holguín-Veras et al., 2016). These technologies align with the circular economy principles, which focus on resource efficiency, minimizing waste, and promoting recycling. The emphasis of Industry 5.0 on environmental sustainability also includes utilizing renewable energy resources along with the creation of intelligent logistics systems that minimize the environmental impact of freight activities (Geissdoerfer et al., 2017). In addition, blockchain technology has the potential to improve transparency as well as traceability across the entire supply chain, aiding in more effective resource management and waste reduction (Saberi et al., 2019). Several case studies highlight the practical applications of Industry 5.0 in urban logistics. DHL's Smart Truck project in Germany, for example, uses IoT and AI to optimize delivery routes in real time, reducing emissions and improving delivery efficiency (DHL, 2017). Similarly, Amazon's implementation of robotics in its fulfillment centers showcases how automation can boost productivity while alleviating physical strain on workers (Ivanov et al., 2017). However, despite the strong potential of Industry 5.0, there are still significant challenges to its widespread adoption in urban logistics. High implementation costs, technological barriers, and a lack of regulatory frameworks are some of the key issues that need to be addressed (Rodrigue et al., 2016). To overcome these barriers, collaboration between governments, private companies, and research institutions is crucial. Public-private partnerships (PPPs) can facilitate the formulation of sustainable urban logistics approaches by leveraging the expertise and resources of the private sector (Kwak et al., 2009a; Zheng et al., 2008). The literature indicates that Industry 5.0 offers substantial opportunities to improve sustainability in urban logistics through the integration of advanced technologies, green practices, and collaborative partnerships. However, realizing these benefits will require overcoming significant challenges related to cost, technology, and regulation.

12.3 RESEARCH GAPS

Although current literature underscores the transformative potential of Industry 5.0 in urban logistics, several critical areas are still underexplored. First, a significant gap exists in empirical studies assessing the practical impacts of Industry 5.0 technologies within urban settings. Most research has focused on pilot programs or specific companies, such as DHL's Smart Truck project (DHL, 2017). Broader studies that assess the scalability and sustainability of Industry 5.0 solutions across diverse urban settings are limited. Future research should focus on conducting empirical evaluations of Industry 5.0 technologies in multiple cities to offer deeper insights into their effectiveness and adaptability. In addition, current literature has not sufficiently explored the influence of government policies in advancing the adoption of Industry 5.0 in urban logistics. Most research emphasizes technological advancements, but little attention is given to the regulatory frameworks needed to support the implementation of Industry 5.0 practices. To fully harness the potential of Industry 5.0, future research should examine how regulatory frameworks can support the adoption of these technologies within urban logistics systems (Rodrigue et al., 2016). Another underexplored area is the social sustainability aspect of Industry 5.0. While much of the literature focuses on the economic and environmental impacts, there is limited discussion on how Industry 5.0 can improve job quality, worker safety, and skill development in logistics. Understanding the social benefits of human-machine collaboration in logistics could provide a more holistic view of Industry 5.0's potential to create safer, more fulfilling work environments (Nahavandi, 2019).

12.4 METHODOLOGY

This chapter employs a theoretical approach to analyze how Industry 5.0 contributes to advancing sustainability within urban logistics. The research draws upon a thorough review of available literature, focusing on the convergence of Industry 5.0 technologies, including AI, IoT, and robotics, and their potential applications in urban logistics. The literature review for this study was conducted by systematically identifying, selecting, and synthesizing research articles, reports, and case studies related to Industry 4.0, Industry 5.0, and urban logistics. Academic databases, such as Scopus and Web of Science, were utilized to gather relevant sources. The selection criteria included peer-reviewed journal articles that discuss the evolution of Industry 5.0, its technological components, and its implications for urban logistics. Keywords used during the search included "Industry 5.0," "sustainable urban logistics," "artificial intelligence in logistics," "IoT in logistics," and "green logistics." The selected literature was evaluated to identify key themes regarding the application of advanced technologies in logistics and their role in promoting sustainability. This chapter also draws on existing theoretical frameworks related to Industry 5.0, sustainability, and urban logistics. The main theoretical concepts explored in this study include the collaborative relationship between humans and machines, which Industry 5.0 promotes, in contrast to the full automation of Industry 4.0. Additionally, the study incorporates green logistics frameworks, examining how technologies can reduce emissions, optimize resource use, and contribute to the circular economy. The chapter also examines the potential of AI, IoT, and blockchain to enhance logistics operations' efficiency, transparency, and sustainability. Through synthesizing these theoretical frameworks and insights from literature, this chapter provides a conceptual analysis of how Industry 5.0 technologies tackle current challenges in urban logistics. The findings and insights from the literature are discussed in the subsequent sections.

12.5 FINDINGS AND DISCUSSION

The findings from this conceptual analysis indicate that Industry 5.0 has the potential to revolutionize urban logistics, driving both operational efficiency and sustainability. Industry 5.0 addresses the intricate challenges of urban logistics, such as last-mile delivery inefficiencies and increasing emissions, by combining human-machine collaboration with advanced technologies like AI, IoT, and blockchain. One of the most significant findings is the importance of human-machine collaboration in urban logistics under Industry 5.0. Human creativity and decision-making become crucial, especially when addressing unpredictable elements like traffic congestion, customer preferences, and real-time disruptions in supply chains. Additionally, adopting Kaizen principles allows logistics providers to implement continuous, incremental improvements. By integrating AI and IoT with Kaizen-driven optimizations, logistics operators can make real-time adjustments that reduce waste, enhance operational efficiency, and drive sustainability (Afthonidis & Tsiotras, 2014). In contrast to the complete automation emphasis of Industry 4.0, the Industry 5.0 model integrates human involvement alongside machines, providing enhanced flexibility and adaptability. AI-driven technologies and continuous improvements through Kaizen principles enable logistics systems to adapt to demand fluctuations, optimize routes, and address traffic congestion in real time, thereby increasing both operational efficiency and sustainability (Bressanelli et al., 2019). As highlighted by Nahavandi (2019), the combination of human intuition with AI's capability to process large datasets leads to a more adaptive and agile logistics system. This finding underscores the notion that, while automation enhances efficiency, the human element is essential for managing the unpredictability inherent in urban logistics. Green logistics technologies play a critical role under the Industry 5.0 structure, particularly in reducing the environmental impacts of urban freight operations (McKinnon et al., 2015). The adoption of EVs and AI-powered route optimization reduces fuel consumption and carbon emissions, especially in last-mile deliveries. IoT-enabled vehicles equipped with sensors allow logistics providers to adjust routes in real time, avoiding traffic bottlenecks and reducing idle times, which leads to significant operational efficiency gains (Morganti & Browne, 2018). The concept of waste elimination (MUDA), which focuses on removing non-value-adding activities, is particularly relevant to urban logistics under Industry 5.0. By reducing unnecessary driving time, optimizing delivery routes, and minimizing idle periods, logistics providers can significantly cut down on emissions and operational costs (Kelp & Becker, 2020). Furthermore, the integration of closed-loop supply chain models that reuse and recycle resources offers a promising pathway for achieving sustainability goals in urban logistics (Kumar et al., 2024). The introduction of UCCs, powered by Industry 5.0 technologies, also represents a significant opportunity for improving urban logistics. UCCs reduce the volume of delivery vehicles entering crowded city centers by consolidating goods at strategic points on the outskirts of cities, where they can then be delivered using fewer vehicles. AI-driven logistics management systems can optimize the timing, routing, and load balancing of these vehicles, ensuring that deliveries are both efficient and environmentally friendly. The DHL Smart Truck initiative in Germany demonstrates how real-time traffic data, AI algorithms, and IoT integration can enhance the effectiveness of UCC operations, resulting in reduced emissions and operational costs (DHL, 2017). By consolidating deliveries and optimizing routes, UCCs can alleviate many of the issues caused by the increasing volume of e-commerce deliveries, such as road congestion, noise pollution, and greenhouse gas emissions. The finding suggests that UCCs, combined with Industry 5.0 technologies, represent a sustainable solution for managing urban freight in dense urban areas.

The role of blockchain technology in improving transparency and accountability across the logistics supply chain is another crucial finding. Blockchain offers a decentralized ledger system that ensures all transactions are recorded and verified, providing transparency at each stage of the supply chain. This transparency is particularly important for logistics providers, as it allows for the monitoring of product journeys from origin to destination, ensuring that ethical and sustainable practices are followed throughout the process (Saberi et al., 2019). Within urban logistics, blockchain technology is able to track deliveries in real time, verify the authenticity of goods, and ensure compliance with environmental regulations. Additionally, blockchain can reduce the risk of fraud and inefficiencies, providing logistics operators with a robust system for managing complex, multi-stakeholder supply chains. The potential for blockchain to facilitate end-to-end visibility and traceability in logistics operations is a key factor in driving greater sustainability and operational efficiency. In addition to blockchain, collaboration between triple helix actors, academic institutions, industries, and governments, fosters innovation and supports the creation of regulatory frameworks necessary for sustainable logistics practices. This collaboration is key in driving research, developing green logistics solutions, and ensuring the integration of Industry 5.0 technologies (Kahn, 2018). Industry 4.0 technologies, such as automation, data-informed decision-making, and digitization, have been transformative in sectors like smart farming, offering increased efficiency, sustainability, and production (Brougham & Haar, 2018). Similarly, in urban logistics, these technologies can enhance operational efficiency by optimizing resource allocation, reducing waste, and improving real-time decision-making through IoT and AI. The adoption of these technologies, supported by regulatory frameworks and incentive programs, is crucial for driving sustainability in logistics, much like their role in agriculture (Batra et al., 2024). Within urban logistics, Industry 5.0 can benefit significantly from the circular economy framework developed through Industry 4.0 technologies like machine learning and IoT. Machine learning techniques help optimize logistics operations by enhancing decision-making processes, such as route planning and resource allocation. It results in reduced emissions, savings in costs, and improved sustainability. IoT devices further support this model by delivering real-time information that tracks vehicle performance and delivery progress, enabling logistics managers to make data-driven adjustments. These technologies, integral to the circular economy, also promote waste reduction through efficient resource management and predictive maintenance, aligning with sustainability goals. In particular, closed-loop supply chain models, which emphasize reusing and recycling resources, offer a promising pathway for achieving sustainability in urban logistics (Bressanelli et al., 2019). By integrating these Industry 4.0-driven techniques with Industry 5.0's focus on human-machine collaboration, logistics providers can significantly enhance both performance and environmental outcomes (Sabale et al., 2024).

In spite of these advancements, the challenges of adopting Industry 5.0 technologies in urban logistics cannot be overlooked. High implementation costs remain one of the most significant barriers, especially for small and medium-sized enterprises (SMEs), who face challenges to invest in the necessary infrastructure and technology. This challenge is compounded by a lack of skilled labor capable of managing and maintaining these advanced systems (Dablanc & Rodrigue, 2014). Future research should focus on creating scalable and cost-effective solutions that enable smaller firms to adopt Industry 5.0 technologies without prohibitive costs. Another critical challenge is the need for policy frameworks to support the widespread adoption of Industry 5.0 technologies in urban logistics. Current regulations often lag behind the pace of technological advancement, and there is a pressing need for governments to create policies that incentivize the use of green logistics technologies and ensure that AI and machine-learning algorithms used in logistics are transparent and accountable. The formulation of PPPs is one avenue for addressing these regulatory gaps. By bringing together governments, private logistics companies, and research institutions, PPPs can foster innovation while also ensuring that the necessary regulatory and financial support is in place (Kwak et al., 2009a). Government involvement in advancing Industry 5.0 in urban logistics is essential not only in terms of providing incentives and subsidies for technology adoption but also in ensuring that these technologies are used ethically and sustainably (Kwak et al., 2009b). Finally, the social dimension of Industry 5.0 in logistics presents both opportunities and challenges. While much of the literature focuses on the economic and environmental benefits of human-machine collaboration, less attention has been paid to how these collaborations impact the workforce.

12.6 CONCLUSION, LIMITATIONS, AND FUTURE RESEARCH AGENDA

This chapter examined how Industry 5.0 could revolutionize urban logistics by combining human-machine collaboration with state-of-the-art technologies like AI, IoT, and blockchain. The findings highlight that Industry 5.0 marks a transition from the automation-based methodology of Industry 4.0 by re-emphasizing human creativity and decision-making in logistics processes. Through this synergy of human and machine efforts, the technologies of Industry 5.0 can effectively tackle many of the sustainability and operational challenges faced by urban logistics, particularly in the areas of last-mile deliveries, emissions reduction, and resource efficiency. The role of green logistics technologies, such as EVs and AI-powered route optimization, has been identified as a critical factor in minimizing the environmental footprint of urban freight systems. UCCs and blockchain technologies further enhance logistics by reducing congestion, improving transparency, and ensuring compliance with environmental regulations. These technologies, when integrated under the Industry 5.0 framework, not only improve the efficiency of logistics operations but also contribute to broader environmental goals, such as reducing carbon emissions and fostering sustainable urban development. This study is theoretical in nature and is based on a comprehensive review of existing literature rather than empirical data. As such, the findings and conclusions are drawn from secondary sources, which may limit their applicability in different real-world contexts. Additionally, while the study highlights several key technologies within Industry 5.0, the implementation challenges for each technology can be different based on factors such as region, scale of operations, and the readiness of stakeholders. These factors were not fully explored in this chapter due to the lack of empirical data. Another limitation of the study is that it emphasizes mainly the technological and operational dimensions of Industry 5.0 in urban logistics, with less emphasis on the socio-political and regulatory dimensions. Future studies should consider these regional differences when assessing the scalability and sustainability of Industry 5.0 solutions in urban logistics. Considering the constraints of this study, several avenues for further study have been identified. First, there is a pressing need for empirical studies to assess the actual impacts of Industry 5.0 technologies in urban logistics. While conceptual frameworks and case studies provide valuable insights, empirical data is necessary to evaluate the effectiveness, scalability, and adaptability of these technologies in diverse urban environments.

REFERENCES

Afthonidis, E. P., & Tsiotras, G. D. (2014). Strategies for Business Excellence under an Economic Crisis. *The TQM Journal*, *26*(6), 610–624. https://doi.org/10.1108/TQM-10-2012-0078

Batra, I., Sharma, C., Malik, A., Sharma, S., Kaswan, M. S., & Garza-Reyes, J. A. (2024). Industrial Revolution and Smart Farming: A Critical Analysis of Research Components in Industry 4.0. *The TQM Journal* (ahead-of-print). https://doi.org/10.1108/TQM-10-2023-0317

Bressanelli, G., Perona, M., & Saccani, N. (2019). Challenges in Supply Chain Redesign for the Circular Economy: A Literature Review and a Multiple Case Study. *International*

Journal of Production Research, *57*(23), 7395–7422. https://doi.org/10.1080/00207543.2018.1542176

Brougham, D., & Haar, J. (2018). Smart Technology, Artificial Intelligence, Robotics, and Algorithms (STARA): Employees' Perceptions of Our Future Workplace. *Journal of Management and Organization*, *24*(2), 239–257). https://doi.org/10.1017/jmo.2016.55

Dablanc, L., & Rodrigue, J.-P. (2014). *The geography of urban freight. In : The geography of urban transportation.*

DHL. (2017). *DHL SmartTruck: Cutting CO_2 emissions with real-time route planning.* https://www.dhl.com/content/dam/dhl/global/csi/documents/pdf/csi-logistics-trend-radar-6-dhl.pdf

Geissdoerfer, M., Savaget, P., Bocken, N. M. P., & Hultink, E. J. (2017). The Circular Economy – A New Sustainability Paradigm? *Journal of Cleaner Production*, *143*, 757–768. https://doi.org/10.1016/J.JCLEPRO.2016.12.048

Holguín-Veras, J., Encarnación, T., Gonzalez-Calderon, C., Winebrake, J., Wang, C., Kyle, S., Herazo-Padilla, N., Kalahasthi, L., Adarme, W., Cantillo, V., Yoshizaki, H., & Garrido, R. (2016). Direct Impacts of Off-Hour Deliveries on Urban Freight Emissions. *Transportation Research Part D: Transport and Environment*, *61*. https://doi.org/10.1016/j.trd.2016.10.013

Ivanov, D., Tsipoulanidis, A., & Schönberger, J. (2017). *Global Supply Chain and Operations Management.* Springer International Publishing. https://doi.org/10.1007/978-3-319-24217-0

Kahn, K. B. (2018). Understanding Innovation. *Business Horizons*, *61*(3), 453–460. https://doi.org/10.1016/J.BUSHOR.2018.01.011

Kaswan, M. S., Chaudhary, R., Garza-Reyes, J. A., & Singh, A. (2024). A Review of Industry 5.0: From Key Facets to a Conceptual Implementation Framework. *International Journal of Quality & Reliability Management* (ahead-of-print). https://doi.org/10.1108/IJQRM-01-2024-0030

Kelp, R., & Becker, H. (2020). *Manufacturing Industries 2030.* https://www.oliverwyman.com/our-expertise/insights/2020/nov/manufacturing-industries-2030.html

Kumar, U., Kaswan, M. S., Kumar, R., Chaudhary, R., Garza-Reyes, J. A., Rathi, R., & Joshi, R. (2024). A Systematic Review of Industry 5.0 from Main Aspects to the Execution Status. *The TQM Journal*, *36*(6), 1526–1549. https://doi.org/10.1108/TQM-06-2023-0183

Kwak, Y., Chih, Y.-Y., & Ibbs, C. (2009a). Towards a Comprehensive Understanding of Public Private Partnerships for Infrastructure Development. *California Management Review*, *51*, 51–78+3.

Kwak, Y. H., Chih, Y., & Ibbs, C. W. (2009b). Towards a Comprehensive Understanding of Public Private Partnerships for Infrastructure Development. *California Management Review*, *51*(2), 51–78. https://doi.org/10.2307/41166480

McKinnon, A. C., Browne, M., Piecyk, M., & Whiteing, A. E. (2015). *Green Logistics: Improving the Environmental Sustainability of Logistics* (3rd ed.). Kogan Page. https://labordoc.ilo.org/discovery/fulldisplay/alma994951688702676/41ILO_INST:41ILO_V1

Morganti, E., & Browne, M. (2018). Technical and Operational Obstacles to the Adoption of Electric Vans in France and the UK: An Operator Perspective. *Transport Policy*, *63*, 90–97. https://doi.org/10.1016/J.TRANPOL.2017.12.010

Nahavandi, S. (2019). Industry 5.0—A Human-Centric Solution. *Sustainability*, *11*(16). https://doi.org/10.3390/su11164371

Rodrigue, J.-P., Comtois, C., & Slack, B. (2016). The Geography of Transport Systems. *The Geography of Transport Systems.* https://doi.org/10.4324/9781315618159

Sabale, D. B., Kaswan, M. S., & Rathi, R. (2024). Development of Circular Economy Model Using Industry 4.0's Machine Learning Techniques to Improve Performance Metrics of MSMEs. *International Journal on Interactive Design and Manufacturing (IJIDeM).* https://doi.org/10.1007/s12008-024-01782-6

Saberi, S., Kouhizadeh, M., Sarkis, J., & Shen, L. (2019). Blockchain Technology and Its Relationships to Sustainable Supply Chain Management. *International Journal of Production Research*, *57*(7), 2117–2135. https://doi.org/10.1080/00207543.2018.1533261

Tao, F., Qi, Q., Liu, A., & Kusiak, A. (2018). Data-Driven Smart Manufacturing. *Journal of Manufacturing Systems*, *48*, 157–169. https://doi.org/10.1016/J.JMSY.2018.01.006

Viu-Roig, M., & Alvarez-Palau, E. J. (2020). The Impact of E-Commerce-Related Last-Mile Logistics on Cities: A Systematic Literature Review. *Sustainability*, *12*(16). https://doi.org/10.3390/su12166492

Zheng, J., Roehrich, J. K., & Lewis, M. A. (2008). The Dynamics of Contractual and Relational Governance: Evidence from Long-Term Public–Private Procurement Arrangements. *Journal of Purchasing and Supply Management*, *14*(1), 43–54. https://doi.org/10.1016/J.PURSUP.2008.01.004

Zhong, R. Y., Xu, X., Klotz, E., & Newman, S. T. (2017). Intelligent Manufacturing in the Context of Industry 4.0: A Review. *Engineering*, *3*(5), 616–630. https://doi.org/10.1016/J.ENG.2017.05.015

13 Towards Sustainable Luxury 5.0

Navigating the Intersection of Development Goals, Luxury Operations and Supply Chains, and Digital Transformation

Hajar Fatorachian

13.1 INTRODUCTION

The luxury industry, traditionally associated with exclusivity and opulence, is undergoing a transformation in response to the growing demand for sustainability and digitalization. Sustainable luxury, a concept aligned with the United Nations' sustainable development goals (SDGs), redefines luxury by embedding social responsibility and environmental stewardship at its core (United Nations, 2015). Simultaneously, the advent of Industry 5.0 technologies, including artificial intelligence (AI), the Internet of things (IoT), and big data analytics, is revolutionizing the way luxury brands operate (Fatorachian and Kazemi, 2021; Schuh et al., 2017). These innovations present new opportunities for enhancing transparency, reducing waste, and meeting the expectations of an increasingly conscientious consumer base (Fatorachian, 2024; Smith and Fatorachian, 2023). This chapter examines how luxury brands can harness Industry 5.0 technologies to adopt sustainable practices while preserving their legacy of exclusivity and craftsmanship (Gavetti et al., 2016; Kapferer and Michaut-Denizeau, 2014).

The luxury industry has long been celebrated for its exclusivity, craftsmanship, and opulence, with these traditional values forming the foundation of its identity. Historically, luxury brands have thrived on principles of scarcity, artisanal quality, and prestige, distinguishing themselves using premium materials and meticulous craftsmanship (Kapferer and Michaut-Denizeau, 2014). However, as the global focus shifts towards sustainable development, the luxury sector is at a pivotal point, balancing its rich heritage with the need for a more responsible and sustainable future (Kering, 2021; Khurana and Ricchetti, 2016).

While luxury's focus on timeless craftsmanship and enduring quality aligns with sustainability to some extent, the industry's environmental and ethical impact has

DOI: 10.1201/9781003535188-13

come under increasing scrutiny. Issues such as resource consumption, waste generation, and carbon footprint have raised concerns that require the luxury sector to reassess its practices (Gardetti and Torres, 2017; Fatorachian and Kazemi, 2018). Moreover, luxury brands are facing pressure not only from consumers but also from regulatory bodies and advocacy groups, all demanding greater transparency and accountability in business operations (Caro and Sadr, 2019; Schuh et al., 2017).

Some scholars argue that luxury's emphasis on durability and craftsmanship inherently supports sustainability principles (Gavetti et al., 2016; Kapferer and Michaut-Denizeau, 2014). However, these qualities alone are insufficient to address the broader challenges of environmental impact and ethical responsibility. The luxury sector must now critically evaluate its role in promoting societal well-being and environmental conservation, responding to these challenges without compromising the essence of its brand identity (Fatorachian, 2024).

13.2 LITERATURE REVIEW

The luxury sector, long associated with artisanal craftsmanship, exclusivity, and opulence, is undergoing a significant transformation with the rise of Industry 5.0. This new era is defined by the integration of cyber-physical systems, advanced AI, the IoT, and big data analytics, which together bring a new level of connectivity, intelligence, and human-machine collaboration to manufacturing and operational processes (Fatorachian, 2023, 2024; Fatorachian and Kazemi, 2021). Industry 5.0 represents a shift from purely automated systems towards a more symbiotic relationship between humans and machines, enabling greater customization, efficiency, and sustainability in production (Kumar et al., 2023; Schuh et al., 2017). As luxury brands embrace these advancements, they are presented with unique opportunities to enhance their operations while meeting the growing consumer demand for responsible business practices. The intersection of digitalization and sustainability offers the luxury sector the chance to maintain its hallmark of opulence and exclusivity while driving innovation in ethical production, transparency, and environmental responsibility (Fatorachian and Smith, 2022; Kapferer and Michaut-Denizeau, 2014; Sabale et al., 2024). This chapter explores how Industry 5.0 is reshaping the luxury landscape and driving the future of sustainable luxury through the adoption of cutting-edge technologies. The integration of smart technologies signals a revolutionary shift in how luxury goods are designed, manufactured, and delivered. Traditionally, luxury brands have relied on manual craftsmanship, with a focus on meticulous, time-intensive processes that preserve the exclusivity and high standards of quality associated with luxury goods. However, Industry 5.0 introduces a new paradigm where real-time data exchange between physical and digital systems enables brands to optimize production in ways that were previously unimaginable (Schuh et al., 2017; Sharma et al., 2023). This transformation allows for more efficient, flexible, and responsive operations, without compromising the artisanal touch that defines luxury products (Rajabzadeh and Fatorachian, 2023). AI-driven algorithms are now capable of enhancing precision in craftsmanship, reducing waste, and minimizing errors—key components in aligning with sustainability objectives (Sabale et al., 2024; Wahab and Nor, 2023). For instance, advanced AI systems can be integrated into the production of high-end fashion, jewellery, or automobiles, where they assist artisans in crafting intricate designs or engineering complex components with a higher degree of accuracy (Kapferer and Michaut-Denizeau, 2014). This real-time feedback from AI systems ensures that human artisans can focus on the creative aspects of their work, while technology manages the repetitive or labour-intensive tasks. The result is a harmonious blend of human expertise and technological precision, which not only elevates the quality of luxury products but also contributes to more sustainable production by reducing material waste and improving resource efficiency (Fatorachian and Kazemi, 2018; Schuh et al., 2017). Moreover, the ability to leverage big data analytics allows luxury brands to predict consumer preferences and adapt their production accordingly (Becker et al., 2017; Roy et al., 2022). By analysing vast amounts of consumer data, AI can offer insights into purchasing patterns, emerging trends, and regional demand, enabling brands to fine-tune their offerings and optimize inventory management. This predictive capability minimizes overproduction, reduces excess inventory, and helps luxury brands maintain their exclusivity while supporting sustainability efforts by avoiding wasteful practices (Fatorachian and Kazemi, 2018, 2021; Gavetti et al., 2016). Enhancing Customization and Consumer Experience.

One of the most profound impacts of Industry 5.0 on the luxury sector is the potential for greater customization and personalization. Historically, luxury has been synonymous with bespoke, tailor-made goods, offering consumers the opportunity to own one-of-a-kind products crafted to their specific preferences (Kapferer and Michaut-Denizeau, 2014). With the advent of Industry 5.0, the ability to deliver personalized products has reached new heights, as luxury brands can now leverage AI, IoT, and big data to gain deeper insights into individual consumer preferences and behaviours (Batra et al., 2024; Fatorachian, 2023; Javornik et al., 2021). This level of personalization not only meets the evolving expectations of modern luxury consumers but also enhances the exclusivity that is central to the luxury experience (Javornik et al., 2021).

Smart technologies enable luxury brands to create highly customized offerings that go beyond traditional bespoke services. For instance, advanced AI algorithms can analyse a customer's past purchases, online browsing behaviour, and even social media interactions to provide highly personalized recommendations for future purchases (Roy et al., 2022). This data-driven approach allows brands to offer unique products tailored to the exact specifications of each customer, from bespoke fashion items to custom-made jewellery or personalized high-end automobiles. Furthermore, the integration of IoT devices within luxury products, such as wearable technology or connected accessories, enables brands to continuously monitor product performance and offer personalized upgrades or services, enhancing the customer's overall experience and extending the lifecycle of luxury goods (Schuh et al., 2017). In addition to offering more personalized products, Industry 5.0 also enables luxury brands to create immersive and interactive shopping experiences. Technologies such as augmented reality (AR) and virtual reality (VR) allow consumers to virtually interact with products, offering them the ability to customize and visualize luxury items in real time (Javornik et al., 2021). For example, a customer shopping for a bespoke watch can use an AR app to visualize different design options, such as the choice of materials, engravings, and finishes, before making a final decision. Similarly, luxury fashion brands can offer virtual fitting rooms where customers can see how a tailored suit or dress will fit their body, all from the comfort of their own home (Kapferer and Michaut-Denizeau, 2014). This level of interaction not only enhances the customer experience but also reduces the need for physical samples, minimizing waste and supporting sustainability efforts by streamlining the production process (Fatorachian and Kazemi, 2021; Roy et al., 2022). The digitalization of the customer experience also fosters a deeper relationship between luxury brands and consumers. By gathering real-time feedback and insights from customers through connected devices and interactive platforms, brands can continuously refine their products and services to meet the evolving preferences of their clientele (Javornik et al., 2021; Kaswan et al., 2024). This two-way communication strengthens brand loyalty and fosters a sense of co-creation, where consumers feel more connected to the products they help design. This level of personalization and engagement is increasingly important in the luxury sector, where consumers are seeking not only high-quality products but also unique experiences that reflect their individual tastes and values (Gavetti et al., 2016).

The integration of Industry 5.0 technologies into luxury supply chains has brought about a revolution in how brands manage their operations, ensuring greater transparency, efficiency, and sustainability (Caro and Sadr, 2019; Fatorachian, 2012). In the past, luxury supply chains were often complex and opaque, with limited visibility into the sourcing of raw materials and the manufacturing processes involved. However, the advent of IoT, blockchain technology, and advanced data analytics has provided luxury brands with the tools to achieve end-to-end visibility across their supply chains, ensuring that every step of the production process is traceable and accountable (Ebinger and Omondi, 2020; Fatorachian, 2024; Kaswan et al., 2024).

Through the use of IoT sensors and real-time tracking systems, luxury brands can monitor the journey of materials from their origin to the final product (Caro and Sadr, 2019). For example, a luxury handbag brand can use IoT-enabled sensors to track the leather used in its products from the moment it is ethically sourced from sustainable farms to its arrival at the manufacturing facility and ultimately to the retail store (Schuh et al., 2017). This level of transparency not only improves operational efficiency by providing real-time data on supply chain performance but also ensures that the materials used in luxury products meet the highest ethical and environmental standards (Kapferer and Michaut-Denizeau, 2014; Smith and Fatorachian, 2023). Blockchain technology further enhances supply chain transparency by creating an immutable digital ledger that records every transaction and movement of goods throughout the supply chain (Mougayar, 2016). This technology allows luxury brands to provide verifiable proof of the authenticity and sustainability of their products, addressing the growing consumer demand for ethically sourced and environmentally friendly goods. For instance, a luxury jewellery brand can use blockchain to trace the provenance of diamonds, ensuring that they are conflict-free and ethically sourced (Narayanan et al., 2017). By providing consumers with access to this information through digital platforms, luxury brands can build trust and demonstrate their commitment to sustainability (Smith and Fatorachian, 2023). The integration of these technologies into luxury supply chains also supports greater flexibility and responsiveness. AI-powered predictive analytics can forecast demand with a high degree of accuracy, enabling luxury brands to adjust their production schedules and inventory levels in real time (Wiedmann et al., 2020). This capability is particularly important in the luxury sector, where the risk of overproduction can lead to excess inventory and waste. By optimizing production based on real-time data, luxury brands can reduce waste, improve resource efficiency, and align their operations with sustainability goals (Becker et al., 2017; Fatorachian, 2023; Schuh et al., 2017).

Luxury brands are increasingly recognizing their responsibility to lead the way in sustainable development, not only because of regulatory pressures but also due to shifting consumer expectations. As consumers become more environmentally conscious and socially aware, they are demanding that the brands they support adhere to ethical and sustainable practices. For luxury brands, this represents both a challenge and an opportunity to redefine what luxury means in the context of the 21st century (Fatorachian and Kazemi, 2021; United Nations, 2015). One of the key areas where luxury brands can make a meaningful impact is in responsible consumption and production, which is central to the United Nations' SDG 12 (Fatorachian, 2024; Kapferer and Michaut-Denizeau, 2014). Luxury brands, with their emphasis on craftsmanship and quality, are well-positioned to champion the principles of sustainability by producing goods that are made to last, reducing the need for frequent replacements, and minimizing the environmental impact of their products. By adopting sustainable materials, ethical sourcing practices, and eco-friendly manufacturing processes, luxury brands can align with global efforts to promote sustainable consumption while maintaining the exclusivity and craftsmanship that define their offerings (Fatorachian and Kazemi, 2021; Gavetti et al., 2016).

Additionally, luxury brands have the potential to influence the broader industry through collaboration and partnerships. Sustainable development requires collective action, and luxury brands, with their global reach and cultural influence, can play a pivotal role in fostering partnerships across industries, governments, and non-governmental organizations (NGOs) to achieve more sustainable outcomes (Fatorachian, 2024; Kaswan et al., 2024; United Nations, 2015). By working together with stakeholders, luxury brands can develop innovative solutions to the challenges of sustainability, from reducing carbon emissions to promoting ethical labour practices in their supply chains (Bocken et al. 2016a, 2016b). While the pursuit of sustainable luxury presents significant opportunities for innovation, it also poses challenges that luxury brands must navigate carefully. One of the primary challenges is balancing the traditional values of exclusivity and opulence with the growing demand for sustainability (Khurana and Ricchetti, 2016). Consumers today are not only seeking high-quality, unique products but are also increasingly concerned about the environmental and social impact of their purchases. Luxury brands must strike a delicate balance between maintaining the allure of exclusivity and meeting the expectations of a more conscientious consumer base (Agrawal et al., 2022). This shift requires luxury brands to reevaluate long-standing industry norms. Traditionally, luxury has been associated with abundance, extravagance, and excess, but sustainable luxury calls for a more nuanced approach—one that values quality, craftsmanship, and environmental responsibility over sheer opulence (Gardetti and Torres, 2017). For instance, brands must rethink their sourcing strategies, opting for materials that are not only rare and valuable but also ethically sourced and environmentally friendly. This redefinition of luxury demands that brands rethink their value propositions and communicate their sustainability efforts in a way that enhances, rather than diminishes, their exclusivity (Brockhaus et al., 2017; Fatorachian, 2013a, 2013b).

At the same time, the shift towards sustainable luxury offers numerous opportunities for brands to differentiate themselves in an increasingly competitive market. By adopting innovative, eco-friendly materials, exploring circular business models, and implementing sustainable packaging initiatives, luxury brands can set new industry standards for responsible business practices. Brands like Stella McCartney, which has pioneered the use of vegetarian leather and sustainable fabrics, and Patagonia, which has embraced a circular business model through its Worn Wear programme, have demonstrated that sustainability and luxury can coexist without compromising on quality or exclusivity (Fletcher, 2019).

13.3 PROPOSED FRAMEWORK AND METHODOLOGY

The methodology for developing this chapter and the proposed framework involved a comprehensive analysis of existing literature and relevant case studies surrounding the integration of Industry 5.0 technologies within the luxury sector. The primary focus was to explore the intersection of advanced technologies—such as AI, big data, the IoT, and blockchain—with sustainable luxury practices. A systematic review of academic articles, industry reports, and case studies from leading luxury brands was conducted to identify key trends, challenges, and opportunities in the field. The aim of this chapter is to explore how Industry 5.0 technologies can drive sustainable innovation in the luxury sector. Specifically, the chapter highlights the role of digitalization in enhancing operational efficiency, improving supply chain transparency, enabling customized consumer experiences, and supporting sustainable practices throughout the value chain. By synthesizing insights from both theoretical and practical sources, the proposed framework offers a structured approach to understanding how luxury brands can align their business models with global sustainability goals while leveraging the capabilities of Industry 5.0.

The framework serves as a guide for luxury brands looking to implement sustainable digital strategies, ensuring that the integration of smart technologies enhances both their environmental impact and their brand value. It also underscores the importance of balancing technological advancement with the preservation of craftsmanship and exclusivity, which remain central to the identity of luxury.

To develop the proposed framework, key themes were first extracted from the literature review. These themes were categorized into main topics like Industry 5.0 technologies and sustainability goals, along with their subcomponents, such as AI, IoT, and outcomes like operational efficiency or traceability. Once the themes were identified, relationships between them were defined, whether hierarchical (parent-child) or associative (cause-effect). After defining the relationships, nodes representing each theme were created and connected by edges to illustrate these relationships. The appropriate diagram type, such as a mind map, hierarchical diagram, or entity-relationship diagram, was selected to reflect the data structure. Finally, the layout and labels were refined to ensure the diagram was clear, readable, and visually well-organized.

The creation of the "Industry 5.0 and Sustainable Luxury Framework" centres on two interdependent pillars: advanced technologies integration and sustainability objectives. The framework highlights how cutting-edge technologies like AI, big data analytics, IoT, and blockchain drive key advancements in the luxury sector. Each technology contributes to specific outcomes—AI enhances operational efficiency, big data enables personalization and customization, IoT promotes supply chain transparency, and blockchain ensures ethical sourcing and traceability. These technological advancements are closely linked to achieving sustainability goals, such as reducing environmental impact, responsible production and consumption, and compliance with global sustainability standards, including alignment with UN SDGs. The framework demonstrates how AI supports reducing environmental impact, while big data aids responsible consumption, and IoT and blockchain ensure compliance with global sustainability targets. These interlinkages create a cycle where technology improves efficiency and transparency while enabling sustainable practices, ensuring the luxury sector remains both innovative and environmentally responsible.

13.4 CONCLUSION

In conclusion, a balanced approach to sustainable luxury within the framework of Industry 5.0 is not only critical but also a strategic necessity. The integration of advanced technologies plays a pivotal role in driving sustainability in the luxury sector. As the industry navigates the dual pressures of digitalization and sustainability, finding harmony between these forces becomes essential to ensuring responsible yet opulent business practices. Industry 5.0 offers luxury brands unparalleled opportunities to align with sustainability objectives. Technologies such as AI, big data, and the IoT significantly enhance transparency, traceability, and environmental performance. These advancements position luxury brands to meet the growing demand for responsible production and consumption, in line with global sustainability goals. However, the luxury sector must recognize that alongside the benefits, digital transformation also presents challenges. Risks related to data privacy, cybersecurity, and the environmental impact of technological infrastructure need careful management. To maximize the benefits of digitalization while safeguarding the core values of luxury, brands must adopt new technologies with caution, ensuring they enhance rather than undermine the exclusive nature of their products.

REFERENCES

Agrawal, G., Rathi, R., Garg, R., and Chhikara, R. (2022). Abundantly rare: Changing consumer trends in the luxury market. Vision.

Batra, I., Sharma, C., Malik, A., Sharma, S., Kaswan, M. S., and Garza-Reyes, J. A. (2024). Industrial revolution and smart farming: A critical analysis of research components in Industry 4.0. The TQM Journal.

Becker, T., Blocher, A., Bücker, I., Drewel, M., Faath, A., Harland, T., Hermann, M., Herzog, G., Hocken, C., Jussen, P., and Krebs, U. (2017). Industrie 4.0 Maturity Index [eng.]: Managing the Digital Transformation of Companies.

Bocken, N. M. P., de Pauw, I., Bakker, C., and van der Grinten, B. (2016a). Product design and business model strategies for a circular economy. Journal of Industrial and Production Engineering, 33(5), 308–320.
Bocken, N. M., Short, S. W., Rana, P., and Evans, S. (2016b). A literature and practice review to develop sustainable business model archetypes. Journal of Cleaner Production, 135, 1215–1230.
Brockhaus, S., Fawcett, S. E., Knemeyer, A. M., and Fawcett, A. M. (2017). Motivations for environmental and social consciousness: Reevaluating the sustainability-based view. Journal of Cleaner Production, 143, 933–947.
Caro, F., and Sadr, R. (2019). The Internet of Things (IoT) in retail: Bridging supply and demand. Business Horizons, 62(1), 47–54.
Ebinger, F., and Omondi, B. (2020). Leveraging digital approaches for transparency in sustainable supply chains: A conceptual paper. Sustainability, 12(15), 6129.
Fatorachian, H. (2012). A critical investigation of electronic supply chain practice among SMEs. International Journal of Advanced Innovations, Thoughts, and Ideas, 1(4).
Fatorachian, H. (2013a). Role of Internet in Supply Chain Integration: Empirical Evidence from Manufacturing SMEs of the UK. In Proceedings of the 9th European Conference on Management Leadership and Governance, ECMLG 2013, Austria.
Fatorachian, H. (2013b). Understanding the Success Factors of Electronic Supply Chain Management: Empirical Evidence from Manufacturing SMEs. In Proceedings of the 16th TOULON-VERONA Conference, 'Excellence in Services', Slovenia.
Fatorachian, H. (2023). The significance of Industry 5.0 in the globalization of supply chain management. European Economic Letters, 13(5), 843. DOI: 10.52783/eel.v13i5.843.
Fatorachian, H. (2024). Sustainable Supply Chain Management and Industry 5.0. In Atiku, S.O., Jeremiah, A., Semente, E. and Boateng, F. (Eds.), Eco-Innovation and Sustainable Development in Industry 5.0. IGI Global. DOI: 10.4018/979-8-3693-2219-2.
Fatorachian, H., and Kazemi, H. (2018). A critical investigation of Industry 4.0 in manufacturing: Theoretical operationalization framework. Production Planning and Control, 29(8), 633–644.
Fatorachian, H., and Kazemi, H. (2021). Impact of Industry 4.0 on supply chain performance. Production Planning and Control, 32(1), 63–81.
Fatorachian, H., and Smith, C. (2022). Impact of CPS on Enhancing Supply Chain Resilience, with a Focus on Solutions to Pandemic Challenges. In Semwal, T. and Faiz, I. (Eds.), Cyber-Physical Systems: Solutions to Pandemic Challenges (pp. 109–125). CRC Press.
Fletcher, K. (2019). Craft of Use: Post-Growth Fashion. Routledge.
Gardetti, M. A., and Torres, A. (Eds.). (2017). Sustainable Luxury: Managing Social and Environmental Performance in Iconic Brands. Routledge.
Gavetti, G., Huang, A., and Rosa, J. A. (2016). Toward a theory of extended contact: The intrinsic benefits of firm-created brand communities. Organization Science, 27(6), 1435–1454.
Javornik, A., Duffy, K., Rokka, J., Scholz, J., Nobbs, K., Motala, A., and Goldenberg, A. (2021). Strategic approaches to augmented reality deployment by luxury brands. Journal of Business Research, 136, 284–292.
Kapferer, J. N., and Michaut-Denizeau, A. (2014). Luxury brand and country of origin effect: Results of an international empirical study. Journal of Business Research, 67(10), 2099–2106.
Kaswan, M. S., Chaudhary, R., Garza-Reyes, J. A., and Singh, A. (2024). A review of Industry 5.0: From key facets to a conceptual implementation framework. International Journal of Quality and Reliability Management (ahead-of-print). https://doi.org/10.1108/IJQRM-01-2024-0030
Kering. (2021). Kering Group: History and Heritage. Retrieved from https://www.kering.com/en/group/history-and-heritage/

Khurana, K., and Ricchetti, M. (2016). Two decades of sustainable supply chain management in the fashion business, an appraisal. Journal of Fashion Marketing and Management, 20(1), 89–104.

Kumar, U., Kaswan, M. S., Kumar, R., Chaudhary, R., Garza-Reyes, J. A., Rathi, R., and Joshi, R. (2023). A systematic review of Industry 5.0 from main aspects to the execution status. The TQM Journal, 36(6), 1526–1549.

Lacity, M. C., Yan, A., and Willcocks, L. P. (2017). A new approach to automating services. MIT Sloan Management Review, 58(1), 23–26.

Mougayar, W. (2016). The Business Blockchain: Promise, Practice, and Application of the Next Internet Technology. John Wiley and Sons.

Narayanan, A., Campani, M., and Hewett, K. (2017). Blockchain for Supply Chain Traceability: Business Requirements and Critical Success Factors. In 2017 IEEE International Conference on RFID Technology and Application (RFID-TA) (pp. 32–39). IEEE.

Rajabzadeh, M., and Fatorachian, H. (2023). Modelling factors influencing IoT adoption: With a focus on agricultural logistics operations. Smart Cities, 6, 3266–3296. DOI: 10.3390/smartcities6060145.

Rejeb, A., Keogh, J. G., and Treiblmaier, H. (2019). Leveraging the internet of things and blockchain technology in supply chain management. Future Internet, 11(7), 161.

Roy, S. K., Singh, G., Hope, M., Nguyen, B., and Harrigan, P. (2022). The Rise of Smart Consumers: Role of Smart Servicescape and Smart Consumer Experience Co-Creation. In The Role of Smart Technologies in Decision Making (pp. 114–147). Routledge.

Sabale, D. B., Kaswan, M. S., and Rathi, R. (2024). Development of circular economy model using Industry 4.0's machine learning techniques to improve performance metrics of MSMEs. International Journal on Interactive Design and Manufacturing (IJIDeM), 1–19.

Schuh, G., Anderl, R., Gausemeier, J., ten Hompel, M., and Wahlster, W. (Eds.). (2017). Industrie 4.0 Maturity Index – Managing the Digital Transformation of Companies. Springer.

Sharma, S., Malik, A., Sharma, C., Batra, I., Kaswan, M. S., and Garza-Reyes, J. A. (2023). Adoption of Industry 4.0 in different sectors: A structural review using natural language processing. International Journal on Interactive Design and Manufacturing (IJIDeM), 1–23.

Smith, C., and Fatorachian, H. (2023). COVID-19 and supply chain disruption management: A behavioural economics perspective and future research direction. Journal of Theoretical Applications and Electronic Commerce Research, 18(4), 2163–2187. DOI: 10.3390/jtaer18040109.

United Nations. (2015). Transforming Our World: The 2030 Agenda for Sustainable Development. New York.

Wahab, N. A. B. A., and Nor, R. B. M. (2023). The role of artificial intelligence in waste reduction in the beverage industry: A comprehensive strategy for enhanced sustainability and efficiency. AI, IoT and the Fourth Industrial Revolution Review, 13(11), 1–8.

Wiedmann, K. P., Hennigs, N., Siebels, A., and Hahner, M. (2018). Measuring consumers' luxury value perception: A cross-cultural framework. Academy of Marketing Science Review, 8(1–2), 1–24.

Wiedmann, T., Lenzen, M., Keyßer, L. T., Steinberger, J. K., and Eisenmenger, N. (2020). Scientists' warning on affluence. Nature Communications, 11(1), 1–10.

Index

Note: Page numbers referring to figures are in **bold** and those referring to tables are in *italic*.

Q

R

S

T

U

V

Kolibri 360 GmbH
Ferdinand-Jühlke-Straße 7
99095 Erfurt, Deutschland
produktsicherheit@kolibri360.de